AMAZONIA

Sussex Library of Study

The Latin America Series

SOCIETY, POLITICS, AND CULTURE

The Latin American Series seeks to establish itself as a venue for the publication of high-quality monographs and edited volumes on Latin American societies, their economies, polities, and cultures. The series is refereed by distinguished scholars in the field. Its editor is Prof. Carlos H. Waisman, Department of Sociology and International Studies Program, University of California, San Diego. Within this series, the Latin American Library, published with the sponsorship of CILAS, the Center for Iberian and Latin American Studies at the University of California, San Diego, is a vehicle for the dissemination of research carried out at that institution.

AMAZONIA

Environment and the Law in Amazonia
A Plurilateral Encounter

James M. Cooper, Co-editor
Christine Hunefeldt, Co-editor
Yesenia Acosta, Assistant Editor

CILAS
CENTER FOR IBERIAN AND LATIN AMERICAN STUDIES
UCSD

SUSSEX
ACADEMIC
PRESS
Brighton • Portland • Toronto

CALIFORNIA WESTERN
SCHOOL OF LAW | San Diego

2 4 6 8 10 9 7 5 3 1

First published in 2013 in Great Britain by
SUSSEX ACADEMIC PRESS
P.O. Box 139
Eastbourne BN24 9BP

and in the United States of America by
SUSSEX ACADEMIC PRESS
920 NE 58th Ave. Suite 300
Portland, Oregon 97213-3786

and in Canada by
SUSSEX ACADEMIC PRESS (CANADA)
8000 Bathurst Street, Unit 1, PO Box 30010, Vaughan, Ontario L4J 0C6

This project was funded by Title VI of the U.S. Department of Education, Proyecto ACCESO, the Avery-Tsui Foundation, the International Legal Studies Program at California Western School of Law, and The Center for Iberian and Latin American Studies and the Institute for International, Comparative and Area Studies at the University of California, San Diego

Cover illustration and design: Marcela Guadiana.
Cover photograph: James M. Cooper.

British Library Cataloguing in Publication Data
A CIP catalogue record for this book is available from the British Library.
Library of Congress Cataloging-in-Publication Data
Environment and the law in Amazonia : a plurilateral encounter / edited by James M. Cooper and Christine Hunefeldt.
p. cm.
Includes bibliographical references and index.
ISBN 978-1-84519-500-7 (h/b : alk. paper)
1. Environmental policy—Amazon River Region. 2. Environmental law—Amazon River Region. 3. Environmental degradation—Amazon River Region. 4. Environmental protection—Amazon River Region. 5. Indigenous peoples—Ecology—Amazon River Region. I. Cooper, J. M. (James M.) II. Hunefeldt, Christine.
GE190.A44E68 2013
333.70981′1—dc23

2012021918

MIX
Paper from responsible sources
FSC® C013056

Typeset and designed by Sussex Academic Press, Brighton & Eastbourne.
Printed by TJ International, Padstow, Cornwall.
This book is printed on acid-free paper.

Contents

Editors' Acknowledgments

We wish to thank our respective institutions for this successful collaboration. We wish to acknowledge the support of University of California San Diego, Dean Seth Lerer, Dean Steven R. Smith of California Western School of Law, Professor David Mares, Director of the Center for Iberian and Latin American Studies, and Tony Grahame at Sussex Academic Press. We thank Nicole Allen for preparing the index. Meighan Sferra, Ruth Padron, and Janice Pezzoli helped coordinate the May 2010 conference.

This project was funded in part by Title VI of the U.S. Department of Education, Proyecto ACCESO, the Avery-Tsui Foundation, and the International Legal Studies Program at California Western School of Law

James Cooper wishes to thank his family. Maybe one day, Jacob will be able to visit a part of the Amazon that has not been cut down.

Christine Hunefeldt thanks her graduate students at UCSD, who unfailingly supported this project all along and, also helped to organize the conference.

Yesenia Acosta wishes to thank her family, and especially her sisters Elisa and Sandra for inspiring a sense of adventure in her at a young age, and setting an example as beautiful, independent women who live life on their own terms.

This book is dedicated to Claes Andreasson who passed away during the final stages of the editing process. Claes, a journalist with Swedish Public Radio, hosted and produced one of the videos on the DVD that accompanies this book and contributed a chapter (*It's the Glacier's Time to Die*). Claes was a committed environmentalist and a soulful professional. He will be missed.

Series Editor's Preface

Carlos Waisman

This volume brings together a series of papers that discuss economic, legal, policy and political issues connected with the Amazon basin. The contributions explore the type of development that evolved in the region, the impact of Amazon deforestation on climate change, international environmental litigation, efforts to control resource development in order to prevent ecological damages, the different ways in which governments manage environmental challenges (or avoid doing so), the consequences of the shift to agriculture and the destruction of ecosystems for indigenous populations, and the mobilization and organization of these communities in order to confront the deterioration of their habitat.

It is difficult to overstate the importance of the Amazon, not only for South America but for the world at large. According to Lawrence Herzog and Richard Finkmoore (papers in this volume), the Amazon rain forest contains 38 percent of the world's tropical biomass, and it produces 20 percent of the world's oxygen. The Amazon river carries 20 percent of the water of all the rivers in the world. Its habitat includes one tenth of all the known biological species. The process of deforestation has been very intense: In the past three decades, an area equal to the size of France has been cleared. Just in the 1990s, deforestation in Brazil has made available for agriculture an area twice as large as that of Portugal. Scientists estimate that, should current trends continue, 40–50 percent of the rain forest will be destroyed in the first half of this century. The ecological damage would be tremendous: Under the most negative of these scenarios, the amount of CO_2 released from now until mid-century would be equivalent to 234 years of the reduction of carbon emission commitments under the Kyoto Protocol.

This process is driven by the growing demand for commodities, not only by established industrial powers but, especially, by China and other new industrial countries in Asia. All the Amazon basin countries have increased dramatically their resource exports in the past couple of decades, but it is in Brazil, which in this period became the world's largest exporter of soybeans, beef, and other commodities, that this transformation has been the most radical.

South American governments seem, in most cases, either irresponsible in their failure to recognize the impending environmental catastrophe, or unable to face the danger in an effective manner. However, when it comes to the environment, and especially when the challenges are long-term (which in politics means "beyond the term in office of the current government"), negation of the issues and collective action paralyses are not the exclusive patrimony of South American governments. We see a similar pattern in those of the United States, most of Western Europe, Japan, and China. Only the prevalence of short-term horizons among policy-makers can explain the inability of the world's governments to co-operate in the control of climate change.

Yet, political institutions do make a difference. Once the public, political elites, and government officials of the industrial countries and China reach the conclusion that concerted action against environmental threats is both necessary and urgent, they will have the organizational and financial resources for designing and implementing effective policies. These are governments whose state capacities (regulatory, extractive, and developmental) are generally strong. The situation is quite different in most of South America.

Whenever academics and public intellectuals within the region analyze governmental passivity or ineffectual response to the looming ecological catastrophe, conspiratorial explanations, which place the blame for this state of affairs on external actors, are standard: the exploitative policies of developed countries, the voracity of multi-national corporations, the "comprador" nature of domestic governments, etc. However, as Ambassador Jeffrey Davidow points out in his afterword, central determinants are the weakness of political institutions, the ineffectiveness of most governments in the region in designing and implementing environmental and economic policies.

State capacities, in particular regulatory, extractive, and developmental capacities, vary a great deal within South America. Of the countries discussed in this volume, Brazil has a state with medium levels of capacity (albeit lower than Chile's, the strongest state in the region), measured in terms of its ability to tax the population, regulate the economy, administer justice, and carry out developmental projects. Peru would have a somehow weaker state, and Venezuela, Ecuador and Bolivia have (together with Paraguay) the weakest states in South America. It is paradoxical that, in this moment, these three countries' governments have adopted populist (i.e. etatist and nationalist) development strategies. These strategies rely on the state, rather than the market, as the main driver of economic activity. State-led development has been very successful in other societies, as the several Asian cases attest, but it looks quite unfeasible in countries whose governments cannot adequately fulfill even their most elementary tasks.

Prospects for an effective response to environmental catastrophe are dire for weak states. For their governments, what is required is not just the awareness of the impending crisis, but also the development of the organizational resources that would allow them to design and implement effective policies.

Academics and public intellectuals could contribute to increase politicians' and, more in general, the public's understanding of these issues. Strengthening the state, making it more autonomous, and focusing its attention on environmental challenges should be at the center of the political agenda. One can only hope that this volume will contribute to this effort.

Introduction

Amazonia: National Politics and Interests, Laws, and the Environment

James M. Cooper and Christine Hunefeldt

What follows are selected contributions from the annual international conference that took place in May 2010 at the Center for Iberian and Latin American Studies (CILAS) at the University of California, San Diego with funding from CILAS' NRC-Title VI and California Western School of Law (CWSL). The conference brought together legal and environmental experts and anthropologists with extensive research experience in the Amazon basin. Unfortunately not all of the nine countries in South America that contain Amazonian lands within their borders could be represented. Our contributions focus on, Bolivia, Brazil, Ecuador, Peru, and Venezuela.

The present volume is divided into three parts. Part I, International Domains, features the contributions of Herzog, Finkmoore, Aceves, and Cooper, the first from San Diego State University and the rest colleagues from CWSL. The aim here is to collect some thoughts pertaining to the international and legal dimensions of the environment and indigenous rights in the Amazon basin. Herzog's introductory essay frames the discussion of sustainable development theory as applied to the Amazon basin. In addition, he provides a succinct description of the Amazon basin in terms of some quantitative global indicators: generation of oxygen, water volumes, bio-diversity. Dependency and dependency theory are in his view still an explanatory scheme, whereby centuries of extractive activities in the region have not led to higher standards of living. To the contrary, poverty for the people living in the Amazon basin has increased, while capital accumulation continues to take place in the "core" areas, namely the already developed and industrialized first world, and the growing urban centers in the Amazon basin itself. The "core–periphery" dynamic, Herzog argues, is observable in the Amazon basin at several levels and with different intensities. In line with what has become a widely accepted insight (at least in the scientific and academic worlds), Herzog proposes to re-assess "development," namely as a combination of productivity, social equality and ecological preservation, what he calls an "eco-development" or, "natural capitalism." Such scientific and academic insights, he argues, need to urgently help toward changing the minds of policy makers, politicians and corporate decision-makers, especially when it comes to such complex mega eco-systems, like the Amazon basin. This should be undertaken in conjunction with efforts undertaken by the United Nations Organization, the United Nations Educational,

Scientific and Cultural Organization, and powerful non-governmental organizations (NGOs), the latter of which are growing in importance.

Finkmoore addresses one specific global issue: climate change and its multiplying projections into the future, particularly through the lens of carbon dioxide emissions; such emissions are a self-perpetuating process by which deforestation (that has occurred in the Amazon basin ever since Spaniards and Portuguese set foot on the continent) releases carbon dioxide increasing planetary warming, further reducing forests after severe droughts, and back to producing more carbon dioxide. A vicious circle, by which presently (2011) one entire soccer field is deforested each minute. Finkmoore provides a fine illustration of the various social and economic reasons that have led to deforestation, especially in Brazil. Partially in answer to Herzog's careful optimism, Finkmoore exposes an international community that is utterly uncommitted to preventing or even reducing deforestation, warming, droughts and the elimination of the Amazon basin in the near future.

Digging further into the international dimensions pertaining to the Amazon basin, with a special eye on indigenous rights, Aceves explores the workings of transnational litigation to protect the Amazon basin, adding to Finkmoore's pessimism about the real possibilities of winning any recognition of indigenous rights vis-a-vis the legal system that has traditionally been controlled by multinational corporations. The primary mechanisms available nowadays to engage in such international disputes are the Alien Tort Statute (ATS), the Torture Victim Protection Act (TVPA), the Anti-Terrorism Act (ATA), and the Foreign Sovereign Immunities Act (FSIA), a cocktail of U.S. legislation that represents a new way of dealing internationally with human rights issues committed outside the United States. These laws allow those wronged by multinational corporations to seek legal damages, what Aceves calls the "plaintiff's diplomacy". For each Act, Aceves provides an empirical case assessment and its past or to be expected outcomes given the workings of these Acts. The main goal of all this legislation is to provide an appropriate mechanism for norm enforcement when corporations violate international law in their overseas operations. Caught up in procedural issues over many years, the effectiveness of such enforcement is still shallow at best, and occurs only after the damage has been done. A pending issue — according to Aceves — is how to translate international law against corporations into domestic legal systems.

The fourth and last contribution to Part I is a discussion of the Yasuní Ishpingo-Tambococha-Tiputini Initiative in Ecuador assessed by author Cooper. In this Initiative, the Ecuadorian government and the United Nations Development Programme (UNDP) have attempted to establish a trust fund to protect the local oil fields from drilling to help save the rain forest in the Amazon basin. This is a new economic model that began to be developed in Latin America in the 1980s, and by which Ecuador would receive some $U.S. 350 million annually for a decade from the international community, an equivalent of what the Ecuadorian government (and not its Amazonian peoples) would have earned from extracting this oil reserve. These are the "debt for nature swaps" that have become part of an international debate and that go hand-in-hand with instrumentalizing international law for the benefit of preservation.

Part II, National Politics and Policies, takes a closer look at what national policies in Venezuela, Bolivia, and Brazil have occasioned recently. What is striking across these four contributions is to see how highly different governmental attitudes and decisions have been over the course of the past years. From the Venezuelan government — as discussed by Yerena and Sartorio — that, although democratically elected — has become an ever more dictatorial regime and has completely obscured transparency when it comes to environmental (especially oil drilling) issues, to an assessment of President Evo Morales' "discourse and deeds" in Bolivia, to the IIRSA projects and the soy moratorium in Brazil as a non-governmental initiative.

The Venezuelan Amazon basin stretches from the south of the Orinoco River to the Amazon River, which also is the international border between Brazil and Venezuela. This stretch of land is inhabited by the largest diversity of Venezuelan indigenous peoples, a fact that is little known. By law, these "backlands" are considered as "empty" public lands. Most of Venezuela's water and energy supplies come from this portion of the country. Hydroelectric power plants have substantially replaced oil consumption, freeing more oil for export. This portion of the country has been placed under 'special administration' terms (ABRAE, Areas Bajo Regimen de Administración Especial). Over time (and during various presidential cycles) there have been shifts between conservationism and conventional development policy, as Yerena and Sartorio tell us. In the meantime, indigenous rights have been completely silenced because of military opposition. As in other parts of Latin America, indigenous peoples have been marginalized to a point where they are now physically located on the very fringes/borders, and as such — in light of their doubtful loyalty to the established state — are positioned as posing a potential threat to the country's territorial and political integrity. The authors perceive this situation as potentially conflict-ridden in the very near future. In Venezuela, unlike elsewhere in the region, the government has been very resilient in adjudicating ancestral lands to its indigenous peoples.

Postero takes us to an actively present Bolivian governmental initiative. In Bolivia, President Evo Morales has launched an international campaign in response to the lukewarm discourse and actions in Copenhagen in 2010, or, in Morales' own words "the so-called developed countries failed in their obligation to provide substantial commitments to reduce greenhouse gases." Global warming in Bolivia has already produced the substantial melting of its life-bringing glaciers (a topic that is further discussed by Andreasson in Part III, where he brings local perceptions to the forefront of his research which enables him to bring in the more cultural components of this environmental threat). According to Postero, Morales' involvement in this international debate has given him a platform for his critiques of imperialism, militarism and neoliberalism. Pollution in Bolivia, on the other hand, has already had a dreadful impact, to the point at which people living nearby mines have seen alterations in their DNA as a result of inhaling toxic air. The Bolivian government has also engaged in redistributing the benefits obtained from the extraction of ores and gas and, seemingly, will do likewise when in the future the huge reserves of lithium will begin to be exported.

Bolivia's other mega project is the Cachuela Esperanza Dam, which is part of the Initiative for the Integration of Regional Infrastructure in South America (IIRSA) projects (discussed in detail for the Brazilian case by Alcantara). Bolivia was, in a sense, born as "primary resource-dependent" and, as Postero argues, it will be very difficult — if not impossible — to alter this growth model. At this point it is still unclear how Bolivia and Bolivians will be able (or not) to solve the dilemma between growing and becoming less poor and protecting its environment and natural resources. What is new, Postero argues, is the "language of contestation", both internally and at the international level.

The dam construction on the Bolivian–Brazilian border stands at the end of the huge and massive Brazilian road construction projects that traverse the entire

Brazilian Amazon basin and are connecting the Atlantic to the Pacific coast and are, as such, crisscrossing many borders and countries on the South American subcontinent. Alcantara, indeed, walks us through the many institutional hurdles the construction of these roads has gone through and the impact the roads have had on the environment dismantling entire bio-ecological systems. The still intensifying development of infrastructure (telecommunications, electricity, and transportation) in the hands of public and private interests are mapped in Alcantara's contribution (BR-364, BR-230, BR-163), as they also show "strategic development hubs" and the very weak and incipient initiative labeled as the "strategic environmental assessment." Alcantara concludes that "development" and "progress" and the ways we think about them are still and mainly profit-driven and in the hands of corporations with little public and local input. If this continues to be the case, the environment, biodiversity, and social welfare are still not an integral part of how we (at least corporations and governments) conceive such urgent considerations as part of a notion of "integral development."

Alcantara's contribution is followed by a contribution by Brown and Koeppe on the recent soy moratorium in Brazil. Brown and Koeppe situate their discussion of the moratorium within a prevailing discussion on how best to deal with environmental and natural resources issues: should such issues be based on state-led regulations and interferences or, should these issues mainly be based on market forces (based in turn on ideas proposed by Fearnside and Nepstad et al., respectively). Resorting to a long-term presentation re conservationism in Brazil (1930–present), Brown and Koeppe conclude based on a thorough analysis of the soy boom in Brazil, that state-led geo-political and historical dynamics played a major role in making a corporate-led moratorium possible. The authors also warn as to the international and national repercussions of nationally engineered biome-centric approaches. It is in their view, very doubtful that globalized markets may have positive conservation outcomes. In other words, a market-driven moratorium implemented by corporations does not stop deforestation and the expansion of the fish-bone grid ever further into the heart of the Amazon basin.

In Part III, Genocide and Plunder, we dive into popular-organized political initiatives in the Andes, mainly in Peru, a country in which mother earth and the environment are nucleating indigenous reactions into a very strong oppositional force. In contrast, Ecuador has witnessed very strong indigenous resistance from the early 1970s, mainly based on claims for land in the highlands; Bolivia, in turn, is witnessing a resurgence of the indigenous rights movement under the aegis of the state, put to use to challenge environmental wisdom as usual (see Postero, herein). The contributions in Part III come from very different angles and perspectives. Varese provides a general introduction to the historical genocide imposed on America's indigenous populations and how neoliberalism is affecting present-day indigenous movements in the Amazon basin, as a new version of the long-term genocidal trend. In Peru's case, Varese argues, indigenous groups in the Amazon have long been trapped between two genocidal national projects: the one imposed by the Shining Path, the other by the army. In both cases, we see nationalistic projects attempting to "assimilate by extermination". Still, Varese argues, the indigenous groups have resisted once again, avoiding the dissolution of their ethnic and communal identities.

Montoya pushes the argument one step further. While reviewing the role of the Left in Peru in understanding Peru's cultural diversity, he argues that recent Peruvian history can be divided into two cycles, one lasting from 1888 to 1980, punctuated by the fight over land, culminating in the guerrilla movement of the 1950s and 1960s, and the agrarian reform implemented by General Juan Velasco Alvarado; and a second cycle that actually began only in 2009, with an upheaval and alliance between Andean and Amazonian indigenous organizations. Such a trend, Montoya argues, would contradict the basic Western Marxist notion that the working class is meant to be the essential moving actor for a revolutionary change in society. There is, Montoya argues, a straightforward compatibility between the worldwide coloniality of power relations and the national dimensions of culture turned into a political subject. Indeed, Montoya goes into great detail to show how Amazonian Indians, having been detached for centuries from national politics, have a different way to do and to engage in politics. Moreover, their demands are both embedded in their own culture (characterized for a deep respect for nature) and coincide with what is proposed at an international scale by environment-reasonable agents.

However, Peru, like all the other countries that share the Amazon basin, does have a government that tends to dismiss cultural arguments that contradict their developmental initiatives. Recently departed Peruvian President Alan García literally declared that his Peruvian brothers in the Amazon basin have not found any fruitful use for the lands they live on and control. Consequently, they should render their ancestral lands for the benefit of all Peruvians and to alleviate poverty. This message encapsulated by a Spanish saying: "El Perro del Hortelano" (*no come ni deja comer*; does not eat and does not allow anyone else to eat) has been portrayed in a documentary film and is explained in Beaulieu's contribution to this volume. The filmmaking process itself was a political and therapeutic act, as many of the affected indigenous groups ended up acting in the film, as did the loosely assembled film crew. The film has had quite an impact in Peru and among indigenous organizations as it drew directly from the events in Bagua, portrayed by Montoya in this volume.

The final contribution by Andreasson — as mentioned earlier — comes back to the "glacier's time to die". His portrayal, however, is a reporter's essay that captures the cultural images on the ground; that is, how people who have had a daily life surrounded by glaciers for endless generations, are now seeing their lives literally "disappear". In Andean cosmology, nature is all-encompassing — it informs their daily lives, their belief systems, their economic, social and political organization. Andreasson manages to bridge local perceptions with the reactions to global warming as portrayed by the scientific community. With the disappearance of natural landmarks, indigenous culture is doomed in the highlands and in the lowlands.

As noted by Ambassador Davidow in his Afterword, the contributions herein leave us with a sense of intimate sadness and a sense of impotence in spite of the depiction of several "good" initiatives, at the international, national, and local levels. However, all these initiatives are still not enough to halt an acceleration of devastation. Be it internal colonialism or international greed, the forces of destruction have gained the upper hand to the detriment of the indigenous peoples in the Amazon basin in the first place, and certainly to the detriment of humanity at large in the not so distant future. If our contributions help heighten a consciousness about what is happening in and with Amazonia, this volume has achieved its main goal. It is our hope — as many authors herein have manifested — that better reason and better understanding may help find better solutions to the long-held dilemma between development, economic growth, and human, social, and biological reproduction.

PART I

International Domains

1 Sustainable Development Theory: Critical Applications to Amazonia

Lawrence A. Herzog

Amazonia is surely the most valuable ecosystem on earth. The Amazon rainforest constitutes about one-third of all rainforests in the world, some 2.3 million square miles. It generates 20 percent of the world's oxygen. Its main waterway, the Amazon River, contains 20 percent of the volume of all rivers on the planet. It may be the most ecologically diverse bio-region on earth, with more than 56,000 species of plants, and thousands of species of birds, amphibians, mammals, and reptiles (da Fonseca and Vogel, 2008). Its plant life contains natural remedies for some of the worst diseases plaguing mankind.

Despite the evolution of technologies that allow for greater monitoring of globally valuable ecological zones, the dangers Amazonia faces from encroachment and economic development (the subject of other chapters in this volume) remain. Sustainable development theory offers a critical blueprint for debates about the future of Amazonia. With this is mind, this article reviews core principles and theories of sustainability and sustainable development, as they apply to the case of Amazonia.

Sustainable development theory evolved as a tool for critically analyzing Western models of growth, which too often proved to be wasteful and inefficient. In cities, industrial-led growth was poorly regulated and generated air and water pollution. In the hinterlands, vast tracts of land, water, and mineral resources were consumed without attention to their long term preservation. Cattle ranches have destroyed tropical rainforests. Croplands in floodplains have been poorly managed. Insecticides, fertilizers, and other chemicals obliterated soils and fragile ecologies. The "modernization" paradigm for development posited the idea that the ills of industrial growth could eventually be mitigated by more and better technology (Rostow, 1960). But, scholars have shown more recently that technology cannot always undo the devastating effects of unsustainable growth. For example, agricultural expansion has depleted soils, while insecticides killed fish stocks near agricultural land (Wheeler, 2004, 19–27).

Amazonia is one of the egregious examples of a larger development pattern that has played out throughout Latin America. Timber production, ranching, and soybean cultivation are wiping out the Amazon rainforest. In Mexico, ranching, which once dominated the northern part of that nation, has begun to sweep across the tropical rainforest regions of the southeast. Over the last 25 years, more than one-fourth of Central America's rainforest has been turned into grass — to feed cattle that supply beef for U.S. hamburger chains (Faber, 2001: 114). Agro-industry throughout Latin America produces animal feed for export, while alienating the bio-region where the agriculture is located. Waste and pesticides are dumped locally; workers are often underpaid, or exploited, and exposed to unhealthy, toxic work environments.

Sadly, Amazonia's plight is yet another example of the legacy of "dependency" in Latin America. "Dependency theory" was a structural explanation of underdevelopment (Cardoso and Faletto, 1979) which ascribed poverty and inequality in less developed nations to a "core/periphery" metropolis/satellite relationship. It argued that poverty is, in fact, systematically maintained by a process of social polarization that makes less advantaged nations (the periphery) dependent on "core" countries. This process operates both on a global scale, and within nations. Within Latin American nations, the "core" took the form of the large metropolis — usually also the national capital — where powerful political officials and economic leaders (plantation owners, investors, bankers, merchants, etc.) lived or had homes. The "periphery" remained in the hinterland — either rural areas where the poor lived, or squatter settlement zones where marginalized rural migrants moved to try to improve their circumstances (Gunder Frank, 1967).

In Amazonia today, we continue to observe a core-periphery dynamic played out on several scales. Within Brazil, the country with the largest amount of Amazonia territory, the Amazonia region's major states (Acre, Amazonas, Para, Rondonia, Roraima, Mato Grosso) remain as a kind of "periphery" — considerably weaker politically than the "core" powerful states of Sao Paulo, Rio de Janeiro, or the southern coastal states of Rio Grande do Sul or Santa Catarina. At the micro level, much of Amazonia's wealth is now concentrated in and around the core metropolis of Manaus, which now contains nearly two million people. It is striking that while Manaus boasts four elegant shopping malls, air conditioned hotels, and foreign-owned assembly plants, much of the rural hinterland of Amazonia continues to be the domain of marginalized indigenous settlements lacking basic health care, nutrition, education or the ability to participate in the political-economic system (Cockburn and Hecht, 1989).

This chapter details some of the ways in which sustainable development theory applies to the specific case of Amazonia. It begins with an overview of the concept of development, reviewing how traditional theories of regional development have evolved to favor a "bottom-up" or grassroots approach. It then tracks various challenges to sustainable development in the Amazonian case, including the problem of the time dimension, how knowledge is organized, and the concepts of limits and of 'place.' It is argued that there are inherent conflicts posed by the increasing globalization of the Amazonian region. However, despite these obstacles, several paths toward sustainable development policy for Amazonia in the future are suggested.

Rethinking "Development": The Sustainability Paradigm and the Amazon

The first task of adopting a sustainable approach to Amazonia is to question the idea of "development." During much of the twentieth century, it was assumed that development was a linear process — growth was projected to occur when labor and capital were applied to natural resources to create production and greater wealth. In traditional economic theory, growth was measured by Gross National Product (GNP), and economists used formulas to measure goods, markets, prices, and scarcities, then determining optimal outputs with different combinations of these variables.

There are at least two major problems with this approach. First, it assumes that all resources are divisible and can be owned. But this really covers only private goods. Much of a nation's resource base, however, lies in its "public goods": water, air, or rainforests — in short, the natural landscape. If nature is not protected — in the form of ecological planning and conservation — a nation's economy and social well-being may suffer in the long term. Second, GNP statistics and linear models of development do not take into account income distribution and inequality.

A sustainable approach, therefore, must revisit the idea of "development." Development must be viewed as a combination of productivity, social equity, and ecological preservation. Nature must become more central to our understanding of capitalism, production, exchange and the meaning of value in a region's economy (Redclift, 1995). We might call this alternative approach "eco-development", economic development tied specifically to cultural/ecological regions and local contexts, thus consistent with the potential of specific regions, with their unique people, geography, and resources (Herzog, 2008).

Eco-development implies that the defining unit of development is based on the rules of nature — that unit is often referred to as a "bio-region". A bio-region is an area defined by nature and culture — a valley, the watershed of a river, a desert basin, a coastal port region, a rainforest.

Eco-development is then measured by planning for a bio-region's "carrying capacity." Carrying capacity as a policy objective demands that elected officials and regional planners assess the natural resource impacts of different human behaviors (economic practices) within a bio-region.

Amazonia is precisely the kind of "mega-bio-region" that screams out for an "eco-development" approach, rather than an "economic development" one. Amazonia's myriad wealth of resources — its biodiversity, its flora and its fauna, the medicinal uses of its plants — have been well documented (Timmons Roberts and Thomas, 2003; da Fonseca and Vogel, 2008). These resources have immense value at both the national and international scales; as such, they demand that politicians and managers adopt an "eco-development" growth strategy in order to preserve them.

More recently, it has been argued that the Amazon has critical value to the world ecosystem and to the prevention of further global climate change (Kirby et al. 2006; World Wildlife Fund, 2011). The Amazon sustains some 60 percent of the world's tropical rainforests. It stores billions of tons of carbon within its biomass and soils; if these biomass resources are erased by deforestation, they have at least two effects. First, when mass vegetation is burned, it emits greenhouse gases, and second, as less rainforest covers the land surface, less moisture is fed back into the atmosphere. This lowers regional and supra-regional humidity, and means less global rainfall, thus further contributing to global warming, as well as to loss of vegetation (Laurence, Powell and Hansen, 2002).

Theories of Development

In the field of development planning, for too long, the dominant paradigms have assumed that growth proceeds outward from the center, and that management of growth must likewise proceed outward from the top of the political hierarchy. This approach has often been termed "development from above" or "top-down" development. Development from above approaches grew out of economic theory and the idea that development could be generated by introducing "growth centers" or industrial development poles into poor regions. The idea was that investing in industrial centers would create jobs and innovation that would "trickle down" to the less advantaged, ultimately transforming poor regions into export-oriented developing zones.

As a subset of top-down development, "growth pole" theory treats industries as the basic unit of analysis, one that exists in an abstract economic space. Economic development is viewed as a structural change brought about by investing in "propulsive industries," growth-inducing factories built in poor regions. According to the theory, these new complexes of industry would spark innovation that would then diffuse to the surrounding region and gradually transform it into a prosperous zone.

In practice, experts have found that development from above often did not work in less developed nations. Innovations introduced in new towns or cities did not necessarily trickle down to indigenous people. Too often growth principles that work in industrialized, wealthy countries do not apply to poor agrarian or developing societies (Friedman and Weaver, 1979; Herzog, 1986). Thus, even when industrial development centers, or "growth poles", are introduced into less developed zones, elite entrepreneurs frequently control the profits, and pay out low wages to indigenous people. Marginal regions end up simply supplying commodities and cheap labor, while the economic power still remains either in the big cities on the coast, or with a small class of entrepreneurs who control development in marginalized regions.

This pattern has played out in Amazonia. The overall development strategy of building up Manaus as a rainforest economic growth pole has had mixed results. Manaus itself is a poorly planned, unsustainable metropolis (Magalhaes and Rojas, 2005). Most of the wealthy residents have moved outside the city, leaving behind an aging downtown whose buildings are falling into disrepair. Favelas have been built along the river and creeks with no attention to controlling the flow of sewage and uncollected garbage. Transportation within the metropolis is dominated by automobiles and trucks; mass transit is poorly articulated (InterAmerican Development Bank, 2010).

Regionally, the river offers an excellent and potentially sustainable transport corridor for Manaus to connect with the hinterland. Yet, the use of the river transit for trade and infrastructure development seems to be overshadowed by the more damaging and unsustainable growth of roads and highways into the interior. And new roads have not offset the reality that much of the indigenous Amazon population continues to remain marginalized from the core.

Amazonia's economic base relies too heavily on either unsustainable industries (like electronic assembly) artificially placed in cities, or unsustainable uses of land in the rain forest that deplete the very resource that could transform Amazonia. "Development from below" or "bottom-up" development offers a more sustainable alternative model. It emphasizes growth tied to local ecology and culture. Development from below recognizes that a successful economic development strategy needs to empower poor, marginalized sub-regions and overcome the centralization of political and economic power that dominates countries like Brazil (Stohr and Taylor, 1981; Weaver, 1984). It further emphasizes the use of appropriate technologies and ecologically sustainable practices that manage the natural environment while allowing for economic development. I give examples of this further below in the final section.

Challenges to Sustainability: The Elements of Time, Knowledge, Limits and Place

Theories of sustainable development require new ways of thinking about the physical environment, space, and place (Wheeler, 2004: 34–40). Let me consider several important themes in sustainability theory, as they apply to a region like Amazonia.In western culture, we tend to emphasize immediacy and instantaneity. Our technologies provide rapid access to information — on computers, cell phones, I-phones, or Blackberries, where we use Google or surf the Internet, or on our televisions where we channel surf. We have developed a sense of entitlement to information. These cyber-electronic machines feed our thirst for speed — they make us expect information to be accessed quickly; we are, therefore becoming impatient with slowness and waiting (Honore, 2004).

This culture of immediacy spills over into the way we govern and into our perception of the politics of the environment. In formal governance circles, there is a tendency to think and operate in short-term frameworks. For example, we make our urban and environmental plans fit into relatively brief time spans — 5, 10, or 20 year units. Meanwhile our elected officials hold offices in even shorter time frames, from 1 to 4 years, with re-election for perhaps a similar amount of time.

Sustainability requires a different mode of thinking. It requires us to understand that the earth changes in long-term cycles — geological shifts, riverbed transformation, watershed evolution, and climate change all occur over massive periods of time. One of the problems of the debate on global climate change, or on the use of sustainable practices, is that politicians and voters are not accustomed to thinking in the long term. One reason for this is that it requires us to put aside our selfish needs in the short term, and consider the future of the planet, and the impact on future generations. In short, we need an environmental-historic approach. This will allow citizens to understand that, as an example, cleaning up the environment can take decades. But not cleaning the environment, or failing to understand the long term dangers of pollution, could permanently scar the planet in irreparable ways.

Time has been a great barrier to the protection of Amazonia. It has taken government officials far too long to craft adequate policies to protect Amazonia. Over the last two or three decades, there has been a creeping acceleration in the decline of ecological integrity in the rain forest. Yet politicians, thinking only within the framework of their political careers and time in office, are often not around to see the long term implications of their work. Meanwhile the rainforest's vulnerable environment becomes further imperiled.

During the 1960s, Brazil's military government initiated pro-development policies that accelerated the pace of deforestation, and environmental destruction of the Amazon. More recently, Brazil has strengthened environmental regulations in

Amazonia, but even progressive president Luiz Inácio Lula da Silva has told the media that outsiders should not criticize Brazil's need to create jobs and profit within the Amazon. This leaves many observers skeptical as to whether the Brazilian government is serious about protecting the long-term sustainability of the rainforest.

Holistic vs. Fragmented Knowledge

Another important distinction must be made between what might be termed "holistic" approaches to environmental planning, as opposed to "fragmented" approaches. In our western culture, we have a tendency to compartmentalize knowledge; we have become an overspecialized society. We train ourselves in very narrow disciplines — law, biology, marketing, — but fail to understand the larger connections between things. Conceptually, this problem is identified as one of "fragmented knowledge." It is sometimes also referred to as "reductionism." We have become a culture of experts who know a lot about one thing, but not enough about larger interconnections. This tendency manifests itself in the environmental and urban planning fields — where we train specialists in oceanography, housing, or land use law, yet not enough scientists who understand the larger picture. This larger picture is often referred to as "holistic knowledge."

We can also see reductionism in the way we create comprehensive urban or environmental plans, or regional plans. These plans tend to be sub-divided into separate categories, or general plan elements — land use, open space, housing, transportation, etc. Policy decisions are then made within each of these 'general plan elements, often without recognizing the interconnections between them.

Fragmentation poses an especially large burden for Amazonia. The region itself is fragmented between different political territories (states) and competing interests (land extraction, beef production, vs. sustainable agriculture). The scale of the region and the diversity of forces at work simply magnify the problem of overspecialization and fragmentation.

Limits

Another important dimension of sustainability is the recognition of limits. In western culture, we tend to make a linear connection between growth and progress. A 1972 book, *The Limits to Growth*, tried to model the consequences of a rapidly expanding world population and finite resource supplies (Meadows, Meadows, Randers and Behrens, 1972). It argued that if we continued to grow at current rates on this planet, there might not be enough resources to house and feed projected future populations.

Western culture is deeply embedded in a period of heightened consumerism, in which citizens are constantly bombarded with subtle and direct messages that consumption is good, the more the better (Lasch, 1979; Sklair, 1991). Our theories of neo-classical economics are based on "growth is good" principles, reduced to concepts such as production, consumption, supply and demand. The fields of economic planning and regional development planning are all based on growth. After a century of unlimited growth, in some parts of the world, we are finally witnessing the introduction of government sanctioned limits — on water use (in desert regions) or on the exploitation of natural resources (in national forests, heavily fished seas).

The question of "limits" is fundamental to the debate about the ecological future of Amazonia. The international scientific community has passionately argued that the Brazilian government needs to limit the construction of highways in the Amazon rainforest, since research shows overwhelmingly that highways are the principal drivers of deforestation (Kirby et al., 2006; Laurence, Albernaz, Fearnside, Vasconcelos and Ferreira, 2004). Highways, it seems, are especially useful to what some have called "the most unsustainable land use in Amazonia — soybean cultivation" (Fearnside, 2001). The Brazilian government counters that scientists from mostly developed nations don't have the right to impose their will on a sovereign nation, especially when the Amazon rainforest development can create jobs and economic development for a nation on the rise. Brazil's government has also argued recently that improvements of frontier governance and environmental law enforcement will enhance the quality and management of Amazonian development. Scientists, however, are unconvinced. In the words of one group of scientists, since 2002, Brazil has still moved forward with the "largest infrastructure expansion (roads, power lines, gas lines, hydroelectric reservoirs, rail, river channelization) in the history of the Amazon" (Laurence, Albernaz et al., 2004).

Place and Sustainability

The emphasis on specialized science and rationality also spills into how we study nature and place. We tend to reduce the natural environment to scientific categories (air, water, land), and then study it with generic land use models. But places are not merely neutral spaces to be broken down into scientific or quantifiable categories — they are unique geographic zones, with their distinct natural landscapes, histories, people, and local character. World cultural regions must protect their uniqueness, in part, because it is how they can develop competitive economies — for example in creating tourism activities, in attracting investors, workers and residents, and in building specialized economies around the mix of site attributes unique to each place.

One of the problems of protecting unique places is that global corporate interests often find it profitable to exploit either their natural resources (as in the Amazon rainforest) or their cultural resources — say, for tourism development. In the case of the latter, too much tourism development can destroy the very thing that tourists come to see. For example, too many cruise ships visiting Antarctica or the Caribbean, or too many eco-hotels and riverboat tours in Amazonia, may compromise the very environments visitors come to see.

Amazonia's value as a place lies in its pristine natural quality, as well as the spiritual and folk value of its indigenous cultures. In the last four or five decades, massive development has begun to compromise Amazonia's powerful sense of place. This can be illustrated using the example of Manaus. Manaus' phenomenal growth since 1960 is driven by its post-rubber boom re-emergence as an economic development region, beginning with the military government's "Brazilian miracle" strategies for the nation in the 1960s. Over the next decade the demographic boom in Manaus reflected the degree to which government investment drew migrants, corporate entrepreneurs and business interests. Manaus' 1960 population of 343,000, doubled to over 600,000 in

1970, and to one million by 1990 (Despres, 1991). That number has again doubled twenty years later to around two million inhabitants.

Manaus has become the urban economic engine for rainforest growth. It is a city that was once known for its quaint post-colonial buildings, highlighted by the iconic Teatro Amazonas, the turn of the century opera building, with a multi-colored dome roof, and neoclassical exterior, and a stunning decorative baroque interior. Today, Manaus can best be described as tropical urban sprawl — a design and environmental nightmare, with very little "sense of place" (Jackson, 1994). The river is congested with chemical plants, oil refineries, electronics assembly plants, and waterfront favelas. A massive exodus to the periphery has turned Manaus into a chaotic, frontier metropolis not unlike Mexican border cities such as Tijuana or Ciudad Juarez — full of crass commercialism and decaying structures in the increasingly abandoned city center. Despite the growth of regional tourism, the city of Manaus continues to have a highly underdeveloped tourism infrastructure. Given the importance of nature to the tourism boom, it is ironic that the hotel industry in Manaus has invested almost nothing in creating green or sustainable ecological designs.

Globalization

Globalization often tends to distort the protection of "bio regions." Because all countries participate in international trade and are part of the world market, as mentioned, national governments have tended to emphasize "growth" over resource conservation, renewable energy, and sustainability. This trend is exacerbated by the role of "foreign aid" — loans, grants, technical assistance, food aid and development projects — which further force nations to favor the creation of capital intensive projects that generate national profit to pay off foreign loans, or to buy goods needed for export-oriented growth. Meanwhile, resources and social needs are neglected (Carothers, 1999; Sachs, 2005; Willis, 2011).

Another important layer of the globalization problem is the "monetization" of local economies. "Middle men" and elite entrepreneurs often control key national economic sectors (large-scale agriculture, commerce, mining), while the poor and working classes often remain trapped in low paying jobs (Redclift, 1995). Meanwhile, national governments try to avoid a debt crisis by using public monies to invest in profitable industry tied to low wages. Not enough public funding is allocated to building sustainable and equitable economies that employ the masses at decent wage levels.

Agriculture is a good example of a local economy that has been globalized to the detriment of poor farmers all over Latin America. "Agro-industry" involves the production, processing, marketing and distribution of food and fiber-products. These products acquire value depending on the global market. Because Latin America and other less developed regions of the world have become dependent on wealthy countries' consumers, they shift their agricultural policies to remain plugged into the buyers' markets in those nations (mainly in Europe, the U.S., Japan, and more recently in India and China).

All of the above forces have crystallized in the Amazonia region. Globalization facilitates a set of "transnational practices" (Sklair, 1991) — which have been imposed on poorer nations and regions like Amazonia — they include free trade agreements, international loans through organizations like the World Bank or the Inter-American Bank, and consumer behaviors promoted through global media. These practices cause further dependency on powerful corporate interests or nations, and reduce local cultures' ability to produce and consume local products within sustainable bio-regions.

Globalizing the Rainforest

Over the last 30 years, national and multi-national corporate interests and the Brazilian government have exploited the rainforest region for short term profit in activities tied to global markets (mainly the sale of beef from cattle ranching, the sale of lumber from logging, and more recently the sale of soybeans), while threatening its long term survival. Some 232,000 square miles of the Amazon have been deforested since 1978. It is a revealing portrait of the problems and politics of unsustainable development.

The most glaring unsustainable practice has been land clearing for cattle ranching. Cattle ranchers clear land to plant grasses to feed herds of beef cattle. The world demand for Brazilian beef has made it a very profitable short-term use of tropical lands in the Amazon (CIFOR, 2010; Kirby et al., 2006) . World demand grew during the 1980s and 1990s when Brazil devalued its currency, thus making beef even more competitive in the world market. Road construction, including the Trans-Amazonian Highway built several decades ago, opened forest lands and made the shipping and packing of beef cheaper.

Equally, the Brazilian government itself has promulgated deforestation indirectly, by sponsoring programs for poor farmers to colonize jungle areas. Unfortunately, the government does not assist farmers in establishing ecologically sustainable best practices for farming. Peasant farmers use fires to clear land, then plant bananas, rice, maize, manioc and palms, which grow well in the short-term, but ultimately deplete soils of their nutrients, thus causing farmers to move on to other lands and repeat the same land clearing and inefficient agricultural practices.

A third and critical factor is infrastructure. The Trans-Amazonian Highway, a federal government rainforest development project from the 1970s, provided for a 2,000-mile highway system that expanded into the massive rainforest to open land for farmers, timber, and mining. The government spurred interest by offering subsidized land, a half year's salary, and other perks. The highway was plagued with problems, including heavy rains, unstable soils, erosion and poor farming yields for settlers. The highway has allowed more access and thus more deforestation as land is cleared for truck farms, cattle, and other production.

Fourth, one cannot underestimate the pull of commercial agriculture. Large scale agri-business arrived in the Amazon in the form of soybean cultivation. Brazilian scientists developed a new variety of soybean that grows well in the rainforest ecosystem. High demand for soybeans, driven by global markets, accelerated soybean production in the jungle.

Finally, there is the logging. The biggest problem with logging is that, although there are laws in the Amazon controlling lumber production, the laws are not sufficiently enforced. There is a great deal of illegal logging in the Brazilian Amazon. Logging is closely tied with road building, so the more the government creates road access into the interior of the Amazon, the more loggers will get in to cut out trees. Those same roads then encourage poor settlers to colonize the interior for temporary agriculture, access to building materials, fuel wood, and hunting game for food. Closer to cities and towns, roads allow favelas and unplanned settlements to expand outward into ecologically sensitive lands.

Sustainable Policy-making: New Ways of Thinking about Amazonia's Future

It can be argued that Amazonia's future development should be subject to barometers of sustainable practice. Consider, for example, the concept of "land ethics". Ethics refers to the relations between individuals and land, and is derived from the larger definition of ethics as "relations between individuals and society" (Leopold, 2004). In Western culture, there is a tendency to over-value private land ownership over the "public interest" (air, water, flora, fauna) (Gore, 2006; Calthorpe, 2010). Western culture placed great value on private property during the nineteenth-century industrial revolution and the economic boom of the twentieth century (De Soto, 2000). In the new millennium we are beginning to realize the danger in ignoring the larger public interest of nature. We need to develop a "land conscience" which would expand the notion of "community" to include the public interest and above all, nature.

In western culture, we also too easily accept the idea that our capitalist market economy should operate like a roller coaster, where profit is best achieved within a cyclical economy, and through boom/bust shifts. This fits the mold of traditional economic theory based on growth. But there are those who argue for a "steady state" approach to economics where governments would regulate the consumption of resources, based on their availability. Since the ecosystem exists in a "steady state" of finite dimensions, it makes sense that the human economy should evolve toward a steady state condition, as opposed to one of limitless growth and unstable booms and busts (Daly, 1973).

Perhaps the most powerful theoretical tool with practical applications to Amazonia is "natural capitalism". Natural capitalism is based on several innovative ideas:

- The environment is not a minor "factor of production," but an envelope containing, provisioning and sustaining the entire economy.
- Economic development should be mediated by "natural capital," that is, it must balance natural resources against the goals of growth and development.
- The loss of natural capital creates wasteful patterns of consumption, population growth and badly designed businesses (for example those that pollute)
- "Resource productivity" must be mediated by human, industrial, financial, and natural capital (Hawken, Lovins and Lovins, 1999).

Conventional capitalism is based on bigger, more efficient factories; it assumes human well-being is achieved by the "highest and best use" of land and that resources should be used to maximize production and profit. It stems from the idea that "environment" is significant only as it balances against economic growth to maintain high standards of living. "Natural capitalism" argues that we can no longer operate our economy under the assumption that we need to build bigger factories, create unlimited growth, drain all land and resources in the interests of maximizing growth, and underestimate the natural environment. On the contrary, "natural capitalism" argues that the environment must become a central factor that drives our economy.

These ideas are critical to the future of Amazonia. The region is inundated by "old school" economic development infrastructure — roads, highways, ports, gas lines, in which the purpose is to maximize temporary growth in lumber, soybean cultivation, and beef production. While these land uses may yield profit in the short and medium term, they are not sustainable in the long term, since they lead to deforestation, which ultimately depletes the region of its self-sustaining natural vegetation, climate system and eco-biological conditions needed for survival and economic sustainability. At some point, scientists have shown, deforestation will significantly endanger rainforest ecology, thus slowing the new economic activities, upon which the Brazilizan (and other) government is counting. Meanwhile, global and local entrepreneurs, as well as governments, have not sufficiently explored the massive potential in alternative economic development sectors, including pharmaceutical, herbal and health benefits from the natural landscape of the rainforest.

One of the main strategies of natural capitalism is what is termed "radical resource productivity," obtaining the same utility (or more) from a product or process while using less material and energy. "Radical resource productivity" suggests technology and creative design can make a product far more ecologically friendly than we currently imagine. This is supported by a second essential element of "natural capitalism" — "bio-mimicry," the idea that industry should be "green" — and more attuned to the natural environment. Instead of building Amazonia's industry around "frontier" strategies of assembly or "free trade zones" (which are footloose and can leave at any time, and are heavy polluters), Amazonia should be concentrating on expanding natural production of biological products taken directly from the local flora. Local products include natural rubber, lumber, Brazil nut, fruits like guarana and acai, and medicinal plants (Mongabay.com). Government officials need to look more closely at these activities, and support agricultural practices that are sustainable, while carefully managing those that are not (lumber and rubber, for example). Further, the economic development potential of activities like medicinal plants must be supported. Greater attention to the use of land for sustainable economic activities could provide incentives to reduce deforestation, which has contributed to global warming. Since 1970, over 230,000 square miles of rainforest has been destroyed (Mongabay.com); this had the dual impact of lowering rainforest capacity to convert carbon dioxide to oxygen (which reduces the greenhouse effect), while increasing burning of the rainforest, which, in turn, adds more carbon dioxide to the atmosphere (Fearnside and Laurence, 2004).

In traditional manufacturing, it has been argued that we have tended to operate with what is termed a "cradle to grave" approach. Products are designed for profit, and with little thought about longevity or about how they impact the environment after they are no longer used. In our traditional industrial culture, products are designed to become obsolete. They move from "cradle" (production in the factory) to "grave" (disposal in a landfill, or incineration), with almost no thought put into sustainable product design. An excellent example is the cell phone. Cell phones are disposed of every few years, as new ones are introduced into the market. Seemingly, because they are so small, cell phones are not necessarily recycled, but rather thrown into enormous disposal yards where billions of cell phone and other old electrical technologies lie.

Cell phone production is one of the important electronic assembly activities in Manaus, Amazonia.

This kind of manufacturing can be termed a "monocultural approach" — one size fits all. We produce homogenous products that are the same everywhere despite different bio-regions. A good example of a monocultural product is the house. Most houses are built in the same manner: First, scrape the site, then construct a house, and finally, landscape with grass and trees, etc. The problem is that different regions require different approaches due to their unique climates, geography, and resources available for building. Yet most developers use "one size fits all" housing prototypes for suburban housing construction all over the world.

An alternative to "cradle to grave" development is what is termed the "cradle to cradle strategy." This approach to manufacturing would: (a) Emphasize use of local material, local geography, and "eco-efficiency"; (b) Use local flora/fauna to craft economic development strategies; (c) Consider natural energy flows: wind towers, solar waves (McDonough and Braungart, 2002). These elements point squarely toward Amazonia, where such sustainable manufacturing seems ideally suited to one of the most potent ecosystems on earth.

To facilitate the transformation of Amazonia's development, international agencies can play an important role. The United Nations Environmental Program is poorly funded and has no jurisdiction over sovereign nations. Non-governmental organizations (NGOs) have stepped up to address environmental issues — for example, the World Wildlife Fund, Greenpeace, and others. The World Conservation Union has stated three ecological strategies that must be followed to make agriculture more sustainable: (a) carefully measure tradeoffs between crop management and cattle ranching; (b) better crop management; (c) protection of watershed forests. Mechanization and the globalization of agriculture are huge threats to preserving local farming.

One hopeful trend is the recognition of "biosphere reserves". A biosphere reserve is an international conservation unit created by the United Nations Educational, Scientific and Cultural Organization (UNESCO) under its program on Man and the Biosphere (MAB). According to "The Statutory Framework of the World Network of Biosphere Reserves," biosphere reserves are created "to promote and demonstrate a balanced relationship between humans and the biosphere." Biosphere reserves serve in some ways as 'living laboratories' for testing out and demonstrating integrated management of land, water and biodiversity. The Amazonian rainforest is one of several biosphere reserves in Brazil, though its designation is mainly for scientific purposes, and thus far, has not yielded to international agreement or monitoring of its effectiveness. One must hope that international development agencies, NGO's, Brazilian planners and environmentalists, and the new frontier of "natural capitalism" entrepreneurs will all ban together and begin to rethink the development strategy for the future of Amazonia.

References

Calthorpe, Peter. 2010. *Urbanism in the Age of Climate Change*. Washington, D.C.: Island Press.

Cardoso, Fernando Enrique, and Enzo Faletto, 1979. *Dependency and Development in Latin America*. Berkeley: University of California Press.

Carothers, Thomas. 1999. *Aiding Democracy Abroad*. Washington, D.C.: Carnegie Endowment.

CIFOR (Center for International Forestry Research). 2010. "Hamburger Connection Fuels Amazon Destruction," www.cifor.org/publications

Cockburn, Alexander, and Susanna Hecht. 1989. *The Fate of the Forest: Developers, Destroyers and Defenders of the Amazon.* New York: HarperCollins.

Da Fonseca, Joao Murilo S. V. and Karen Vogel. 2008. "Law and Policy in Brazil: Protecting the Rainforest and Enhancing Communities," unpublished paper, Midwest Political Science Association, April, Chicago. www.allacademic.com

Daly, Herman. 1973. "The Steady State Economy," from Herman Daly, *Towards a Steady State Economy.* San Francisco: W.H.Freeman.

De Soto, Hernando. 2000. *The Mystery of Capital.* New York: Basic Books.

Despres, Leo. 1991. *Manaus: Social Life and Work in Brazil's Free Trade Zone,* Albany, N.Y: SUNY Press,

Faber, Daniel. 2001. "Revolution in the Rainforest," in Susan E. Place, ed. *Tropical Rainforest: Latin America, Nature and Society in Transition,* pp. 97–122. Wilmington, DE: Scholarly Resources.

Fearnside, Philip M. 2001. "Soybean cultivation as a threat to the environment in Brazil." *Environmental Conservation,* Vol. 28, No. 1, 23–38.

Fearnside, Philip M. and William F. Laurence. 2004. "Tropical forestation and greenhouse gas emissions," *Ecological Applications,* Vol. 14, 982–86.

Friedman, John and Clyde Weaver. 1979. *Territory and Function: The Evolution of Regional Planning.* Berkeley: University of California Press.

Gore, Al. 2006. *Earth in the Balance.* New York: Rodale.

Gunder Frank, Andre. 1967. *Capitalism and Underdevelopment in Latin America.* New York: Monthly Review Press.

Hawken, Paul, Amory Lovins and L. Hunter Lovins. 1999. *Natural Capitalism,* New York: Little, Brown and Co.

Herzog, Lawrence. 2008. *Nature and World Cultures.* San Jose: Center for Global Studies, San Jose State University.

Herzog, Lawrence A. 1986. "The integrated spatial development model and regional planning in the Peruvian Andes: a critique," *International Review of Administrative Sciences,* Vol. 52, 301–324.

Honore, Carl. 2004. *In Praise of Slowness.* San Francisco: HarperCollins.

InterAmerican Development Bank. 2010. "Manaus: A Case Study." http://idbdocs.iadb.org/wsdocs/getdocument.aspx?docnum=2011606

Jackson, John Brinckerhoff. 1994. *A Sense of Place, a Sense of Time.* New Haven: Yale University Press.

Kirby, Kathryn, William F. Laurance, Ana K. Albernaz, Gotz Schroth, Philip M. Fearnside, Scott Bergen, Eduardo M. Venticinque, and Carlos da Costa. 2006. "The Future of Deforestation in the Brazilian Amazon,"*Futures,* Vol. 38, 432–453.

Lasch, Christopher. 1979. *The Culture of Narcissism.* New York: Norton.

Laurence, William F., George Powell and Lara Hansen. 2002. "A Precarious Future for Amazonia", *Trends in Ecology and Evolution,* Vol. 17, No. 5, June, 251–252.

Laurence, William F., Ana Albernaz, Philip Fearnside, Heraldo Vasconcelos and Leandro Ferreira, 2004. "Deforestation in Amazonia," *Science,* Vol. 34, May, 1109.

Leopold, Aldo. 2004. "The Land Ethic", A Sand County Almanac, 1949. In Stephen W. Wheeler and Timothy Beatley, eds. *The Sustainable Urban Development Reader.* New York: Routledge, 2004, 20–29.

Magalhaes, Fernanda, and Eduardo Rojas. 2007. "Facing the challenges of informal settlements in urban centers: The re-urbanization of Manaus, Brazil," Inter American Development Bank, *Working Paper.* http://idbdocs.iadb.org/wsdocs/getdocument.aspx?docnum= 1442373

Mc Donough, William and Michael Braungart. 2002. *Cradle to Cradle: Remaking the Way We Make Things.* N.Y.: North Point Press.

Meadows, Donella H, Dennis L. Meadows, Jørgen Randers, and William W. Behrens III. 1972. *The Limits to Growth.* New York: Universe Books.
Mongabay.com, 2011. "Saving the Amazon rainforest." www.mongabay.com
Redclift, Michael. 1995. *Sustainable Development.* New York: Routledge.
Rostow, Waft W. 1960. *The Stages of Economic Growth.* Cambridge: Cambridge University Press.
Sachs, Jeffrey. 2005. *The End of Poverty.* New York: Penguin.
Sklair, Leslie. 1991. *Sociology of the Global System.* Baltimore: Johns Hopkins University Press.
Stohr, W.B. and D.R. Taylor, eds. 1981. *Development From Above or Below?* New York: John Wiley & Sons.
Timmons Roberts, J. and Nikki Demetria Thomas, eds. 2003. *Trouble in Paradise: Globalization and Environmental Crisis in Latin America.* New York: Routledge.
Weaver, Clyde. 1984. *Regional Development and the Local Community.* New York: Wiley and Sons.
Wheeler, Stephen M. 2004. *Planning for Sustainability.* New York: Routledge.
Willis, Katie. 2011. *Theories and Practices of Development.* New York: Routledge.
World Wildlife Fund. 2011. "Safeguarding the Amazon." www.worldwildlife.org

2 Critical Connections: The Amazon and Global Climate Disruption

Richard Finkmoore

Amazonia is a place of environmental superlatives. It contains the world's largest watershed drained by the world's longest, deepest, and widest river.[1] With more than one thousand tributaries, the Amazon River discharges into the Atlantic Ocean about fifteen percent of all freshwater in the world.[2] Amazonia supports an extraordinary concentration of biological diversity. Scientists estimate that one in ten known species in the world are found there;[3] species as yet undiscovered may add another ten percent to this total.[4] A third of the known vascular plants are found in Brazil, Colombia, and Peru. Brazil alone has more than fifty thousand species of animals and plants.[5] With an indigenous population of 17.6 million, Amazonia also supports significant human cultural diversity.[6] It is home to about 380 — and maybe as many as 500 — distinct groups of peoples,[7] speaking more than 650 different languages and dialects.[8] And, most importantly for this chapter, Amazonia encompasses the largest remaining area of tropical rainforest in the world.[9]

Today, this unique biome faces unprecedented threats. The rainforests of Amazonia, once considered too vast and too impenetrable to succumb to human exploitation, are now endangered by human-caused global change in the form of climate disruption.[10] In general terms, forests and climate are connected by a positive feedback loop: deforestation releases vast amounts of carbon dioxide (CO_2) into the atmosphere; this increases planetary warming, which further reduces forests, in part, through increasingly severe droughts, which lead to even more CO_2 being released when drought-stricken trees die. It is a sequence that may become self-perpetuating.

The details of these critical connections are the subject of the present chapter. It begins by describing the role of tropical forests in the global carbon cycle and then turns to the contribution of forest lost generally to climate change. The chapter then examines the causes of deforestation in Amazonia, its consequences for the world's climate, and the effects of climate disruption on the Amazon rainforest. Because sixty percent of the Amazon rainforest is located in Brazil,[11] this chapter emphasizes conditions in that country. The discussion leads to a stark reality that demands unparalleled efforts to protect both the Amazon forest and to stabilize the global climate system.

Roles of Forests in the Planetary Carbon Cycle

The world's forests — particularly its tropical forests — are intimately related to our changing climate. Although not the most potent greenhouse gas, CO_2 is by far the most important, since it comprises 50 to 60 percent of all greenhouse gas emissions and has a long atmospheric lifetime. Forests play a major role in the short-term planetary carbon cycle, acting as a vast carbon sink and reservoir[12] holding almost half of the world's terrestrial carbon.[13] Most of this carbon is contained in living trees and other vegetation. Forested areas contain about 75 percent of all carbon stored in plants and hold about forty percent of all carbon stored in soil.[14] One researcher estimates that forests store more carbon than all the world's oil reserves.[15]

Forests cover about 30 percent of the earth's surface and tropical forests comprise approximately half of that amount. Yet the vegetation and soils in tropical forests hold as much carbon as all temperate-zone and boreal forests combined. Moreover, trees in tropical forests contain about 50 percent more carbon than the same area of trees in non-tropical forests.[16] In addition to their indispensable role as carbon reservoirs, growing forests continually sequester carbon through photosynthesis.[17] In the case of tropical forests, this process removes huge amounts of carbon from the atmosphere, perhaps as much as 20 percent of the annual global greenhouse gas emissions.[18] Thus, tropical forests play a particularly critical role in the global carbon cycle and, as discussed below, continued human interference with this cycle is likely to have dramatic consequences for climatic stability.

Tropical Forest Loss and Climate Change

Deforestation is the overwhelming threat to tropical forests throughout the world. Trees on approximately thirteen million hectares (a hectare is approximately 2.5 acres) of all forest types are removed each year. Most of the current deforestation takes place in the tropical developing countries where the deforestation rate is about three times higher than the global rate. In the fifteen years from 1992 to 2007, an alarming 8.3 percent of all tropical forest area was lost.[19]

The agents of deforestation range from transnational corporations in search of profits, to indigenous people seeking merely to subsist. Loggers (who may be landless peasants or industrial timber companies), miners, large commercial farmers typically producing soybeans or palm oil, cattle ranchers, and fuel wood collectors, are all contributing agents of deforestation.[20] Frequently, these parties interact in the process of deforestation in the following manner:

> For example, cattle ranchers may obtain access to lands by deforesting but also by inducing landless peasants to do the job for them. With the possibility of having their land legally acquired through occupation, and subsequent sale of property to cattle ranchers, landless peasants can be effective — and dependent — agents of deforestation. Some logging companies are known to supply local populations with power saws and then buy their production of logs, often obtained illegally from protected areas.[21]

The indirect drivers of forest loss are numerous and powerful. Misdirected government policies that provide subsidies, encourage the concentration of ownership, or require the improvement of forest lands to obtain land tenure rights, are a few examples.[22] The inadequate enforcement of existing forest laws and corruption also play a role. National governments often build highways into remote forested areas to promote economic development. Denise de Alcantara's chapter in this book explains such initiatives by the Brazilian government.

Deforestation has many negative effects, such as the reduction of biological diversity and displacement of indigenous people who depend on "intact" rainforests. Another major consequence is that the carbon stored in the forest is released into the atmosphere. This occurs immediately if trees are burned or slowly by the decay of other organic matter. Only a fraction of the carbon initially stored in a forest becomes lumber that is used in building long-lasting structures.[23] Most forest vegetation ends up as waste which oxidizes and releases CO_2 to the atmosphere. On a worldwide basis, deforestation and related land-use activities emit up to an estimated 5.9 GtC (gigatons of carbon) which is the equivalent of about 20 percent of global CO_2 emissions from all sources.[24] (For a comparison, all fossil fuel use and cement-making release approximately 6.4 GtC per year.)[25] Emissions from deforestation alone — almost all of which occurs in the tropics — comprise a larger share of CO_2 emissions (17 percent) than the entire global transportation sector (13 percent).[26]

According to the Fourth Assessment Report of the authoritative Intergovernmental Panel on Climate Change (IPCC), curbing deforestation is "the [emission] mitigation option with the largest and most immediate carbon stock impact in the short term."[27] Moreover, reducing deforestation is considered a highly cost-effective way of reducing greenhouse gas emissions. However, the most significant lesson is that the potential contribution of greenhouse gases to the atmosphere from tropical forest degradation means the emissions reductions necessary to avoid the worst consequences of climate disruption can be achieved only by dramatic decreases of the loss of tropical forests.[28] In the words of the Interacademy Panel on International Issues, "There can be no solution to climate change without addressing deforestation."[29]

Deforestation in Amazonia

Forest loss in the Amazon is extensive. About 17 percent of the original expanse of the forest or sixty million hectares has been cleared in the last three decades,[30] equaling an area the size of France by 1997.[31] In Brazil, location of the largest share of the Amazon forest, the area of accumulated deforestation is 15 percent.[32] Deforestation there has been described by Professor Lesley McAllister as a "drama" with a large "cast of characters."

> "[P]rimary roles [are] played (in rough order of appearance) by land grabbers who claim land using fraud and violence; loggers who extract the most valuable species from the land; colonists and other subsistence farmers who buy or simply occupy land; and capitalized farmers and large cattle ranchers who often buy land from the land grabbers, colonists, and subsistence farmers. Supporting roles are played by gold miners and money launderers, as well as migrant laborers who sometimes become debt slaves to land grabbers, farmers, and ranchers."[33]

Thus, the deforestation of Amazonia is the result of a complex interaction of many factors, including demand for timber, subsidies for agribusiness, uncertainties in land tenure rights, and investment in large-scale infrastructure projects.[34]

In Brazil, an "arc of deforestation" runs through the country's eastern and southern forests.[35] Annual rates of deforestation — as determined by satellite imagery by Brazil's National Institute for Space Research (INPE) since 1988 — have fluctuated considerably. Between 1990 and 2000, deforestation in the Brazilian Amazon measured 17.2 million hectares, an area twice the size of Portugal.[36] Forest loss in Brazil peaked in 1994–95, then dropped noticeably, but rose again in 1997–98, then climbed dramatically in 2002.[37] According to government data, rates then fell for three consecutive years in the early 2000s — but during this time forest still disappeared at a rate of about four football fields per minute.[38] Record low losses in Brazilian forests occurred from August 2008 to July 2010, the lowest in 22 years.[39] Thus, in April 2011, the Norwegian environment minister could point to "[t]he big success story [. . .] Brazil, where deforestation has been reduced by 70 percent in seven years."[40]

But later, new satellite images showed deforestation increased almost six-fold in a single year from August 2010 to April 2011. The Brazilian environment minister called the figures "alarming."[41] In late May 2011, Brazil announced the creation of a "crisis cabinet" to "suffocate environmental crimes." The Institute of Environment and Renewable Natural Resources (IBAMA), the country's environmental protection agency, said it would launch 200 law enforcement operations in the region by the end of the current year.[42] Some environmentalists believe that underlying the dramatic increase in deforestation are proposed changes to Brazil's Forest Code, under active consideration by the legislature as of the writing of this chapter.[43] In late May 2011, Brazil's lower house of Congress overwhelmingly passed a bill which would strengthen forest protection in some respects, but also includes amnesty provisions that would suspend fines and replanting requirements for landowners who own parcels of up to four hundred hectares that were illegally cut before July 22, 2008.[44] (On that date, a presidential decree imposed additional requirements on the recovery of illegally cut land). Even though this legislation would not grant amnesty to current violators, if enacted, the law arguably would create the expectation of more amnesty in the future.[45] The 2010–11 increase in deforestation is reminiscent of the sharp rise in forest clearing in Brazil during the last half of 2007, which "dashed hopes that recent conservation initiatives and law enforcement efforts have finally wrested control of the rain forest from development interests."[46]

In just the last quarter century, several potent factors emerged that jeopardize the Brazilian Amazon rainforest. The principal threats are beef production, soybean cultivation, and to a lesser extent, population pressure and infrastructure development. In the late 1990s, commodity production in the Amazon overtook other causes as the primary reason for forest loss. Enterprise-driven deforestation largely supplanted poverty-driven deforestation.[47] Until 1991, Amazonia produced beef only for domestic consumption, but Brazilian meat consumption per capita was rising rapidly, quadrupling between 1972 and 1997.[48] After 1994, production expanded so rapidly that Brazil became the world's largest beef exporter.[49] This was accomplished by the large-scale conversion of forest to pastures, because between 1990 and 2002 most of the growth in Brazil's livestock population was in the Amazon region.[50]

The factors driving Brazilian beef exports are varied. One influence was the devaluation of the national currency between 1998 and 2002. This effectively doubled the price of beef at home, creating a significant incentive for ranchers to expand production. Furthermore, the price of Brazilian beef on the international market fell, making the country's exports more competitive.[51] Also, the substantial eradication of foot-and-mouth disease within Brazil's cattle industry removed a bar to beef exports. Very low land prices in the Amazon region — partly the result of ineffective law enforcement against those who illegally occupy government land– also contribute to making ranching profitable.[52]

Beginning in the 1990s, soybean agriculture in the Amazon expanded as well, but it is responsible for just one-tenth of the area of forest converted to pasture. In Brazil, most soy is grown in savanna and other non-rainforest areas, but in the five years from 1999 to 2004, soy cultivation grew by 15 percent in the Amazon rainforest.[53] Most of the deforestation for soybean cultivation is indirect: as land prices rise, cattle ranchers sell to soy farmers and then move on to other areas on the frontier where yet more forest is cleared for cattle raising.[54]

The growing worldwide demand for biofuels could create pressure for additional forest clearing in Amazonia.[55] Soybean oil can be used to make biodiesel fuels which may well lead to the expansion of soy cultivation in the Amazon. In part, this would be a consequence of United States' energy policy, which continues to provide generous subsidies for corn ethanol. This has discouraged American farmers from growing soybeans, which encourages the growing of soy in Brazil.[56] Sugar cane ethanol also is in demand because it is one of the most efficient and inexpensive forms of ethanol, and Brazil is its largest exporter, with plans to double its production. Thus, there is significant incentive to expand production of sugar cane in the country. While most of this is expected to occur in southern and central Brazil, it will likely displace soybean production and cattle ranching from those areas, creating additional pressure for even more commodity production and further land clearing in the rainforest to the north.[57]

The development of Amazonia has been made possible by the construction of roads and other infrastructure in the region. Investments in roads connecting the central and southern parts of Brazil to the rainforests in the 1960s initiated deforestation there.[58] There have been large private and governmental investments made for storage and processing facilities, development of a barge system, construction of deep-water ports, and the paving of all-weather highways.[59] These projects have made it possible to move agricultural crops from the Amazon to distant markets. For example, the Transoceanic Highway, which links the central Amazon to the Pacific Ocean ports in Peru, now provides an additional route for exports to Asia. One stimulus to agricultural expansion into the rainforests is the $43 billion *Avanca Brasil* (Forward Brazil) program for the construction of various types of infrastructure, such as gas and power lines, hydroelectric facilities, and railroads.[60]

The rise of Brazil as an agro-industrial giant — now the world's largest exporter of soybeans as well as beef — means that external factors have come to have considerable influence on deforestation rates. For example, the fate of its forests is now tightly linked to world commodities markets.[61] The increasing demand for meat in the rapidly developing countries of the world, such as China, raises prices for beef, which incentivizes deforestation in Brazil. The global financial crisis of 2008–9 contributed to a sharp decline in commodity prices generally and the near elimination of credit for deforestation activities, causing a drop in the rate of forest loss.[62] Some observers also attribute the slowdown to campaigns by activist groups such as Greenpeace; these groups have targeted major companies which buy leather from ranchers involved in Amazon deforestation.

Climate Consequences of Amazon Deforestation

As the world's largest remaining area of tropical forest, the Amazon contains about 38 percent of all tropical biomass. The forest is a vast reservoir holding 140 Pg (petagrams) of carbon, the equivalent of 436,000 million tons of CO_2.[63] Scientists have estimated this amount to be the equivalent of between nine and fourteen years of current annual global carbon emissions.[64] One study has found that without changes in government policy, at least one-half of the Amazon rainforest will be destroyed or damaged by 2030;[65] another has predicted that 40 percent of the existing Amazon forest will be eliminated under business-as-usual scenarios by 2050.[66]

In either case, a tremendous amount of carbon will be unleashed. Under the 2050 scenario, about 117 billion tons of CO_2 would be released. Other research, using conservative estimates, concludes that between fifteen and twenty-six billion tons of carbon will be released to the atmosphere through early-stage forest die back.[67] These amounts dwarf the hoped for — but unrealized — reduction of carbon emissions by two billion tons under the Kyoto Protocol during the 2008–2012 commitment period.

Effects of Climate Change on the Amazon Rainforest

While deforestation is the most apparent change to the forests of Amazonia, human-induced climate disruption has emerged as the greatest threat to their future. The Interacademy Panel on International Issues has concluded that the natural carbon capture and storage function of the world's undisturbed tropical forests — their value as carbon sinks — "is already being compromised by climate change."[68] This is equally true for the Amazon. As one scientist has declared, "the climate has already become dangerous" for the Amazon rainforest.[69] The major regional impacts of climate disruption are a lengthened dry season and an increase in the frequency of droughts. In the Amazon region, reduced precipitation and higher temperatures could result in a die back of over three-quarters of its forested areas.[70] This would occur if the rainforest reaches a critical "tipping point," although the process may take several decades to unfold.[71]

When considering predictions, it should be noted that there is still considerable uncertainty about the impacts of climate change on Amazonia.[72] Computer modeling of such impacts has improved, but some models reach different or inconsistent results. Nevertheless, there is a broad consensus among recent studies, if not in the details of their forecasts, sufficient to paint a bleak picture of the Amazon rainforest's long-term future.

The first forecast of dramatic impacts on the Amazon forests from global climate change was produced in 2000 by a computer model at the Hadley Centre in the U.K. Meteorological Office.[73] These results were possible because the Hadley model incorporated feedback effects which had not been included in previous climate modeling. Since then, other climate models have incorporated such amplifying factors, so that now a substantial majority of models predict reduced rainfall and increased temperatures in Amazonia.[74] The decline in rainfall is anticipated to exceed

20 percent in the eastern Amazon forests by the end of this century.[75] More recently, the Hadley Centre model has indicated that the region will experience very large increases in temperature, from 8.7 to 14.0 degrees C. by 2100. Even at the lower end of this range, the consequences for the rainforest "would [. . .] be a catastrophe."[76] One result would be the large-scale replacement of rainforest with savannas and semi-arid vegetation.

A November 2009 report by the Tyndall Centre for Climate Change Research reviewed the growing number of climate studies concerning Amazonia.[77] These analyses found that the region is subjected to drying during El Niño years and when the North Atlantic Ocean is unusually warm. For example, the extreme drought of 2005 in the western and southern areas of the Amazon has been linked to warm sea-surface temperatures in the North Atlantic. Recent studies based on anticipated atmospheric concentrations of greenhouse gases predict that droughts comparable to the 2005 event will become much more common in Amazonia. The 2005 drought was an approximately one-in-twenty year occurrence (other studies suggest it was a one-in-one hundred year event). Such droughts are forecast to become a one in every other year event by 2025 and a nine in every ten years occurrence by 2060.[78]

In fact, the 2005 drought was followed just five years later by an even more severe dry spell. In 2010, the Amazon River at Manaus, Brazil, fell to its lowest level in half a century, and several Amazon tributaries dried up almost completely.[79] The earlier drought caused the Amazon rainforest to release five billion tons of CO_2 from tree mortality and decay. (By comparison, all fossil fuel use in the United States — the world's second largest emitter — added 5.4 billion tons of CO_2 in 2009.) The more recent drought caused precipitation decline over a three million square kilometer area of forest, compared with 1.9 million square kilometers in 2005. An initial estimate has shown that the 2010 drought likely released more CO_2 than the earlier event.[80] In view of these numbers, an important recent study has concluded:

> "The two recent Amazon droughts demonstrate a mechanism by which remaining intact tropical forests of South America can shift from buffering the increase in atmospheric carbon dioxide to accelerating it. Indeed, two major droughts in a decade may have largely offset the net gains. [. . .] Thus, repeated droughts may have important decadal-scale impacts on the global carbon cycle.
>
> [I]nteractions among climatic changes, human actions, and forest responses, represent potential positive feedbacks that could lead to widespread Amazon forest degradation or loss."[81]

The Tyndall Centre report also reviewed a number of studies which forecast a widespread die back of the Amazon rainforest as droughts become more frequent. Perhaps as much as 70 percent of the forest would be lost due to climate change, and such a die back would be essentially irreversible.[82] Notably, one recent study suggests that the Amazon rainforest "could be committed to long-term die back long before any response is observable."[83]

One major change in Amazonia over the last fifty years has been the forest's increasing susceptibility to fire.[84] Not surprisingly, this phenomenon is exacerbated by droughts caused by climate change. Intact Amazon forests are remarkably resistant to fire because the fuel layer on the forest floor must become dry enough to burn; this level of drying is rare in mature tropical forests because the air is exceptionally humid.

In large parts of Amazonia, selective logging, deforestation, drought, and fires are thinning the forest canopy, which permits more sunlight to penetrate to the fuel layer on the ground. As increasing numbers of trees are killed, the leaf canopy is opened and the fuel on the ground becomes increasingly dry, which is the major determinant of forest flammability.[85]

Because the green leaf crowns of the rainforest canopy trees take many years to regrow, the drought-caused deaths of trees may well increase fire susceptibility for years afterwards. Thus, with every new fire the forest becomes more susceptible to subsequent fires. A scientist with more than twenty-five years of work in the Amazon, Daniel Nepstad, has identified an "ecological tipping point" which is reached when the Amazon forests become so flammable that frequent burning is virtually inevitable.[86] Furthermore, the loss of forests through deforestation or climate change-caused die back leads to less evapotranspiration which in turn, reduces the supply of water vapor needed to maintain high rainfall levels. This forms another positive feedback leading to more fires and more forest loss.[87]

Confirmation of such grim predictions is found in another Hadley Centre computer modeling study released in March 2009 in which scientists concluded that up to 85 percent of the Amazon rainforest could be lost if worldwide greenhouse gas emissions are not brought under control.[88] Even under moderate climate change scenarios, severe forest loss is likely to occur. The study found that a two-degree C. rise in average temperature above pre-industrial levels — increasingly thought to be an optimistic target — would cause an Amazon die back of between twenty and 40 percent within one hundred years. Moreover, the concept of committed ecosystem change strongly suggests that at any point in time, the rainforest will exhibit only a portion of the eventual die back: "For example, by 2050 when die back begins to be observed in the simulation, the forest is already committed to eventually losing 50% of its area even without further increases [in climate change]."[89] Summarizing the study results, one of the Hadley researchers said, "the impacts of climate change on the Amazon are much worse than we thought."[90]

Conclusion

Once known for its unparalleled ecological attributes, the Amazon rainforest "has become a powerful symbol of human impact on the planet."[91] As the forest is leveled to produce agricultural commodities for the world, and droughts and fires increase, the synergistic impact of the multiple threats becomes greater and clearer. Referring to the prospect for damage to the forest from climate change, Peter Cox, a professor of climate system dynamics, observed: "Ecologically it would be a catastrophe [. . .]. The tropics are drivers of the world's weather systems and killing the Amazon is likely to change them forever." An honest appraisal of the situation leads to this stark reality: preserving Amazonia's forest is necessary to stabilize the world's climate at relatively safe temperatures; and stabilization of the world's climate at safe temperatures is necessary to preserve the Amazon rainforest.[92] Accomplishing both will require massive changes on the ground in Amazonia and in humanity's discharges of greenhouse gas pollutants to the atmosphere. The international community's disappointing efforts on both fronts thus far should be cause for alarm.

Notes

1 Manuel Picasso Botto, *The Amazon Cooperation Treaty: A mechanism for cooperation and sustainable development, in* MANAGEMENT OF LATIN AMERICAN RIVER BASINS: AMAZON, PLATA, AND SAO FRANCISCO 68, 69 (Asit K. Biswas et al. eds. 1999).
2 *Id.*
3 Ahmed Djoghlaf, Executive Secretary, Convention of Biological Diversity, Message of the Executive Secretary on the Occasion of the Technical Workshop on Protected Areas in the Amazon 2 (United Nations Environment Programme 2008).
4 TYNDALL CENTER FOR CLIMATE CHANGE RESEARCH, MAJOR TIPPING POINTS IN THE EARTH'S CLIMATE SYSTEM AND CONSEQUENCES FOR THE INSURANCE SECTOR 54 (2009) [hereinafter TYNDALL CENTER].
5 UNITED NATIONS ENVIRONMENT PROGRAMME & AMAZON COOPERATION TREATY ORGANIZATION, ENVIRONMENTAL OUTLOOK IN AMAZONIA 112 (2009) [hereinafter GEO AMAZONIA].
6 Botto, *supra* note 1, at 69.
7 *Id.*
8 GEO AMAZONIA, *supra* note 5, at 21
9 TYNDALL CENTER, *supra* note 4, at 53.
10 John P. Holdren, *Global Climate Disruption: What Do We Know? What Should We Do?* John F. Kennedy Jr. Forum, Harvard University (Nov. 6, 2007) available at http://belfercenter.ksg.harvard.edu/publication/17661/global_climate_disruption.html ("'Global warming' is a misnomer because it implies something gradual, uniform, & benign, none of which is true; 'global climatic disruption' is a more accurate description.").
11 PAULO BARRETO ET AL., HUMAN PRESSURE ON THE BRAZILIAN AMAZON FORESTS 3 (World Resources Institute 2006).
12 CHRIS WOLD, DAVID HUNTER & MELISSA POWERS, CLIMATE CHANGE AND THE LAW 8 (2009).
13 R. A. Houghton, *Tropical Deforestation as a Source of Greenhouse Gas Emissions,* in TROPICAL DEFORESTATION AND CLIMATE CHANGE 14–15 (Paulo Moutinho and Stephan Schwartzman, eds. 2005).
14 FOOD & AGRICULTURE ORGANIZATION OF THE U.N., GLOBAL FOREST RESOURCES ASSESSMENT 2005: PROGRESS TOWARDS SUSTAINABLE FOREST MANAGEMENT xii (2005).
15 ERIN C. MYERS MADERIA, POLICIES TO REDUCE EMISSIONS FROM DEFORESTATION AND DEGRADATION (REDD) IN DEVELOPING COUNTRIES 18 (Resources for the Future 2008).
16 Houghton, *supra* note 13 at 15.
17 DAVID HUNTER, JAMES SALZMAN & DURWOOD ZAELKE, INTERNATIONAL ENVIRONMENTAL LAW AND POLICY 619 (4th ed. 2011).
18 William Boyd, *Ways of Seeing in Environmental Law: How Deforestation Became an Object of Climate Governance,* 37 ECOLOGY L.Q. 843, 867–868 (2010).
19 MYERS MADIERA, *supra* note 15, at 18.
20 ARNOLDO CONTRERAS-HERMOSILLA, THE UNDERLYING CAUSES OF FOREST DECLINE 4 (Center for International Forestry Research 2000).
21 *Id.* at 6.
22 *Id.* at 10–16.
23 Houghton, *supra* note 13, at 13.
24 MYERS MADERIA, *supra* note 15, at 19.
25 *Id.*
26 Houghton, *supra* note 13, at 18.
27 UNITED NATIONS FRAMEWORK CONVENTION ON CLIMATE CHANGE, FACT SHEET:

REDUCING EMISSIONS FROM DEFORESTATION IN DEVELOPING COUNTRIES: APPROACHES TO STIMULATE ACTION 1 (2009).

28 UNION OF CONCERNED SCIENTISTS, U.S. POLICY AND PREVENTING TROPICAL DEFORESTATION, available at http://www.ucsusa.org/global_warming/solutions/forest_solutions/us-policy (last visited May 2, 2010).

29 INTERACADEMY PANEL ON INTERNATIONAL ISSUES, IAP STATEMENT ON TROPICAL FORESTS AND CLIMATE CHANGE 1 (2009).

30 Boyd, *supra* note 18, at 867.

31 Philip M. Fearnside, *Deforestation Impacts, Environmental Services and the International Community*, in AMAZONIA AT THE CROSSROADS: THE CHALLENGE OF SUSTAINABLE DEVELOPMENT 11 (Anthony Hall, ed., 2000).

32 Lesley K. McAllister, *Sustainable Consumption Governance in the Amazon*, 38 ENVTL. L. RPTR.10873, 10874 (2008).

33 *Id.*

34 PAULO MOUTINHO, TROPICAL FOREST, DEFORESTATION AND CLIMATE CHANGE: THE AMAZON CASE 2–3 (Heinrich Boll Stiftung n.d.).

35 Ane Alencar et al., *Carbon Emissions Associated with Forest Fires in Brazil*, in TROPICAL DEFORESTATION AND CLIMATE CHANGE 23 (Paulo Moutinho and Stephan Schwartzman eds. 2005).

36 DAVID KAIMOWITZ ET AL., HAMBURGER CONNECTION FUELS AMAZON DESTRUCTION: CATTLE RANCHING AND DEFORESTATION IN BRAZIL'S AMAZON 2 (Center for International Forestry Research 2004).

37 *Id.*

38 World Wildlife Fund, *Update: Falling Amazon Deforestation Rates Create Opportunity for Other Damaged Forests*, available at http://www.worldwildlife.org/who/media/press/2009 1-2- (last visited Dec. 16, 2009).

39 Environmental News Service, *Brazil Protects Amazon With Record Low Amazon Deforestation*, available at http://www.ens-newswire.com/ens/dec2010/2010-12-01-01.html (last visited Dec. 2, 2010).

40 Environmental News Service, *WWF Marks 50 Years With Urgent Call for Green Economy*, http://www.ens-newswire.com/ens/apr2011/2011-04-29-01.html (last visited April 30, 2011).

41 BBC News Latin America & Caribbean, *Brazil: Amazon Rainforest Deforestation Rises Sharply*, available at http://www.bbc.co.uk/news/world-latin-america May 19, 2011) (last visited May 22, 2011).

42 environmentalresearchweb, *Brazil Forms 'Crisis Cabinet' Following Unexpected Deforestation Surge*, available at http://environmentalresearchweb.org/cws/article/news46083 (May 25, 2011) (last visited June 1, 2011).

43 BBC News, *supra* note 41.

44 *Brazil's Lower House Approves Bill to Provide Forest Code, With Amnesty for Past Logging*, 34 INT'L. ENVT. RPTR. 563 (June 8, 2011).

45 *Id.*

46 Rhett Butler, *Global Commodities Boom Fuels New Assault on Amazon*, YALE ENVIRONMENT 360 1, available at http://e360.yale.edu/content/print.msp?id=2010 (2010).

47 Rhett Butler, *Rainforest Conservation: a Year in Review*, available at http://news.mongabay.com/2009/1228-rainforests.html (December 27, 2009) (last visited Apr. 28, 2010).

48 Kaimowitz et al., *supra* note 36, at 3.

49 McAllister, *supra* note 32, at 10875.

50 Kaimowitz et al., *supra* note 36, at 2.

51 *Id.* at 3–4.

52 *Id.* at 5.
53 McAllister, *supra* note 32, at 10875.
54 McAllister, *supra* note 32, at 10876.
55 *Id.*
56 Butler, *supra* note 46, at 2.
57 Daniel C. Nepstad, The Amazon's Vicious Cycles: Drought and Fire in the Greenhouse 13 (WWF 2007).
58 Moutinho, *supra* note 34, at 3.
59 McAllister, *supra* note 32, at 10875.
60 Butler, *supra* note 46, at 3.
61 McAllister, *supra* note 32, at 10880.
62 Butler, *supra* note 47.
63 Tyndall Center, *supra* note 4, at 53.
64 Nepstad, *supra* note 57, at 7.
65 McAllister, *supra* note 32, at 10874.
66 Boyd, *supra* note 18, at 867.
67 Nepstad, *supra* note 57, at 15.
68 Interacademy Panel on International Issues, *supra* note 29, at 1.
69 Philip M. Fearnside, *Tropical Forests of Amazonia*, in Climate Change Science and Policy 105 (Stephen H. Schneider et al. eds. 2010).
70 I. Allison et al., The Copenhagen Diagnosis: Updating the World on the Latest Climate Science 43 (2009).
71 *Id.*
72 University of Leeds (U.K.), *Two Severe Amazon Droughts in Five Years Alarm Scientists*, available at http://www.leeds.ac.uk/article.1466 (Feb. 11, 2011) (last visited June 8, 2011).
73 Fearnside, *supra* note 69, at 105.
74 *Id.* at 106.
75 Nepstad, *supra* note 57, at 12 (Fig. 3).
76 Fearnside, *supra* note 69, at 106.
77 Tyndall Center, *supra* note 4, at 19–20, 82.
78 *Id.* at 19.
79 BBC News Latin America & Caribbean, *Severe Drought Affects Brazilian Amazon*, http://www.bbc.co.uk/news/world-latin-america-11610382 (Oct. 22, 2010) (last visited Jan. 20, 2011).
80 University of Leeds, *supra* note 72.
81 Simon L. Lewis et al., *The 2010 Amazon Drought*, 331 Science 554 (Feb. 4, 2011).
82 Chris Jones et al., *Committed Terrestrial Ecosystem Changes Due to Climate Change*, 2 Nature Geoscience 484, 486 (July 2009).
83 *Id.* at 484.
84 Nepstad, *supra* note 57, at 8.
85 *Id.* at 9.
86 *Id.* at 8.
87 Fearnside, *supra* note 69, at 108.
88 David Adam, *Amazon Could Shrink by 85% Due to Climate Change, Scientists Say*, The Guardian (UK), March 11, 2009, available at http://www.guardian.co.uk/environment/2009/mar/11/amazon-global-warming-trees/print (describing Chris Jones et al., *supra* note 82).
89 Chris Jones et al., *supra* note 82, at 484.
90 *Id.*
91 *Id.*
92 Nepstad, *supra* note 57, at 4.

3 Using Transnational Litigation to Protect the Amazon: A Study on Corporate Liability

William J. Aceves

Introduction

Transnational litigation uses domestic litigation to address injuries that occurred in other countries.[1] It most often addresses violations of domestic law. In the United States, it has been used with increasing frequency to address violations of international law committed in other parts of the world.[2] In recent years, multinational corporations have been prominent subjects in this litigation, and defendants have included Chevron, Chiquita, Coca-Cola, Drummond, Exxon Mobil, Firestone, KBR, Occidental Petroleum, Pfizer, Royal Dutch, Shell, Unocal, and Yahoo. Corporate activity in the developing world has been subject to particular scrutiny.[3]

For several reasons, corporations have been subjects of transnational litigation.[4] First, corporations control significant resources, which translate into economic wealth and political influence. These resources allow corporations to wield enormous power, particularly in developing countries. Indeed, public-private partnerships allow corporations to wield state-like power in many countries. Second, many corporations are transnational in scope, which makes them more difficult to regulate. The transnational nature of the corporate structure makes regulatory oversight exceedingly complex. Corporate governance rules differ widely throughout the world. Environmental standards and labor practices are also distinct. In fact, there are profound differences between corporate, environmental, and labor standards in developed and developing countries. In developing countries, such standards are often minimal or non-existent. Third, corporate projects can involve multi-billion dollar ventures that affect thousands of people, from factory workers to individuals displaced by extraction operations. The consequences of corporate activity can, therefore, be significant. For these reasons, there is growing scrutiny of corporate activity abroad.[5] Indeed, the corporate responsibility movement has grown significantly in recent years in response to these developments.[6]

There are also several reasons why these lawsuits have been filed in the United States.[7] The U.S. legal system is more receptive to tort litigation than other countries. There are extensive procedural mechanisms available to plaintiffs, including broad-based discovery rules. In addition, there is the potential for significant financial redress, particularly through punitive damage awards. At the same time, foreign jurisdictions where the harms occurred are often unwilling or unable to hear these cases. In some countries, the legal systems are not well developed and lack adequate resources for trying complex lawsuits. In other countries, the governments may be implicated in the underlying harms and may block efforts to litigate these claims. When human rights abuses or environmental harms occur, therefore, transnational litigation may be the sole mechanism for redress available to victims.

For these reasons, it should not be surprising to recognize the role of transnational litigation in addressing corporate activity in the Amazon region. The large scale of the Amazon region, the enormous resources found throughout the region, and the lack of effective regulatory oversight have all led to extensive corporate activity and its attendant harms.[8] In the absence of effective domestic recourse, local groups in the Amazon region have sought redress for environmental harms and human rights abuses through litigation in U.S. courts.[9] In *Aguinda v. Texaco*, for example, a group of Kichwa Indians filed a class action lawsuit against Texaco in federal district court for environmental harms and associated health problems caused by oil extraction operations in Ecuador.[10] In *Carijano v. Occidental Petroleum*, a group of Achuar Indians sued Occidental Petroleum for causing extensive environmental harm in Peru.[11] Both lawsuits were filed in U.S. courts in an effort to address harms resulting from corporate activity in the Amazon region.

This chapter examines how transnational litigation has been used in the United States to address environmental harms perpetrated by multinational corporations in the Amazon region. It provides a brief overview on transnational litigation and how it functions in the United States and reviews how this litigation has addressed claims against corporations for their overseas operations. Finally, it examines how transnational litigation has been used to address environmental harms in the Amazon region by reviewing both *Aguinda v. Texaco* and *Carijano v. Occidental Petroleum* as prime examples of such litigation.

A Brief History of Transnational Litigation

In an era of extensive interdependence and globalization, disputes often transcend national borders. But the concept of transnational litigation, where lawsuits can address matters across borders, has existed for centuries.[12] "Civil actions in tort have long been considered transitory, in that the tortfeasor's wrongful act created an obligation to make reparations that followed him across national boundaries and was enforceable wherever he was found."[13]

Transnational litigation most commonly applies to domestic claims.[14] For example, violations of U.S. law can be litigated in U.S. courts even if the underlying actions occurred abroad. While the possible extraterritorial application of U.S. law has been limited in recent years by the U.S. Supreme Court, it remains viable in appropriate circumstances.[15] In addition to U.S. claims, violations of foreign law can also be litigated in U.S. courts even if the underlying actions occurred abroad. The concept of the transitory tort allows for injured plaintiffs to sue defendants even in jurisdictions far removed from the place of injury.[16] Of course, the defendants must be subject to service of process and personal jurisdiction in all these cases.

Transnational litigation involving international law claims is a distinctly American phenomenon, which traces its origins to the First Judiciary Act of 1789.[17] The ATS is codified at 28 U.S.C. §1350 and now provides that "[t]he district courts shall have original jurisdiction of any civil action by an alien for a tort only, committed in violation of the law of nations or a treaty of the United States."[18] Thus, the ATS has three elements: (1) an alien; (2) sues for a tort; (3) alleging a violation of the law of nations or a treaty of the United States. While the ATS was adopted in 1789, the statute remained dormant for almost 200 years and there were few references in the case law.[19]

In 1979, a Paraguayan family brought a lawsuit in federal court in New York against a Paraguayan police official who was living in the United States.[20] Federal jurisdiction was premised on the ATS. The plaintiffs alleged that the defendant, Americo Pena-Irala, had tortured and killed their family member, Joelito Filartiga, in Paraguay. The lawsuit asserted that torture was a tort, and that torture was a violation of international law. The federal district court originally dismissed the lawsuit, holding international law did not prohibit acts of torture perpetrated by a country against its own citizens.[21] The Second Circuit Court of Appeals overturned this decision in 1980.

In *Filartiga v. Pena-Irala,* the Second Circuit held that an alleged torturer could be sued under the ATS.[22] The court reviewed extensive international practice to determine that torture was a clear violation of international law. The court also found the ATS authorized a legitimate exercise of jurisdiction by the federal courts. Thus, "whenever an alleged torturer is found and served with process by an alien within our borders, [the ATS] provides federal jurisdiction."[23] According to the court, the decision offered "a small but important step in the fulfillment of the ageless dream to free all people from brutal violence."[24] On remand, Pena-Irala defaulted, and the district court entered a judgment in favor of the Filartiga family, awarding them $10 million in damages.

Since *Filartiga* was decided by the Second Circuit in 1980, there have been an increasing number of ATS cases filed in U.S. courts. Plaintiffs have come from Argentina, Bosnia, Burma, Chile, Ecuador, El Salvador, Ethiopia, Haiti, Honduras, Indonesia, Nigeria, and Somalia.[25] Alleged torts include torture, genocide, war crimes, crimes against humanity, summary execution, forced disappearance, sexual violence, slavery, and forced labor.[26] Defendants have included foreign police officers, military officials, private individuals, and multinational corporations.[27]

In 2004, the U.S. Supreme Court reaffirmed the *Filartiga* decision in *Sosa v. Alvarez-Machain,* holding the Alien Tort Statute established federal jurisdiction for certain cases involving international law violations.[28] In *Sosa,* a Mexican doctor sued several individuals for their role in kidnapping him from Mexico and bringing him to the United States. The complaint alleged several causes of action, including arbitrary detention. In *Sosa,* the Supreme Court held that federal courts can recognize private causes of action for certain torts in violation of international law (also known as the law of nations). The Court noted, however, that only certain claims may be pursued. "[C]ourts should require any claim based on the present-day law of nations to rest on a norm of international character accepted by the civilized world and defined with a specificity comparable to the features of the eighteenth-century paradigms we have recognized."[29] Thus, the Court identified two key elements necessary for identifying viable claims under the ATS: (1) international norms that are accepted by the civilized world; and (2) defined with specificity.[30]

The Torture Victim Protection Act (TVPA) provides another mechanism for transnational litigation in the United States.[31] Congress adopted the TVPA in 1991 to supplement the remedies available under the ATS. The TVPA establishes a federal cause of action for acts of torture and extrajudicial killing committed abroad.[32] The statute provides, in pertinent part, that "[a]n individual who, under actual or apparent authority, or color of law, of any foreign nation, (1) subjects an individual to torture shall, in a civil action, be liable for damages to that individual; or (2) subjects an individual to extrajudicial killing shall, in a civil action, be liable for damages to the individual's legal representative, or to any person who may be a claimant in an action for wrongful death."[33] The TVPA differs from the ATS in several respects. The ATS is a jurisdictional grant and allows courts to establish a cause of action for serious violations of international law whereas the TVPA solely establishes a cause of action. In addition, the ATS is only available to foreign nationals whereas the TVPA is available to both U.S. citizens and foreign nationals. Finally, the TVPA contains several procedural refinements regarding exhaustion of domestic remedies and statutes of limitation.[34]

The ATS and TVPA are the primary mechanisms for transnational litigation involving international law claims in the United States. Other mechanisms for transnational litigation include the Anti-Terrorism Act and the Foreign Sovereign Immunities Act.[35]

Transnational Litigation Involving Corporate Defendants

There has been a growing movement to use transnational litigation against corporations for actions arising out of corporate activity abroad. Some of these lawsuits allege corporate complicity for human rights violations. Other cases assert environmental claims.

One of the earliest examples of transnational litigation involving a corporation was *Doe v. Unocal*.[36] In 1991, the Unocal Corporation entered negotiations with the Burmese government to develop a natural gas field and pipeline in Burma (Myanmar). Unocal would develop the pipeline, and the Burmese government would provide security and labor. During the development of the project, the Burmese government committed numerous human rights abuses, including compelling local villagers to provide forced labor; torturing and killing villagers who opposed the project; forcing relocation of villagers; and taking property, including crops and livestock, from villagers. In October 1996, several Burmese villagers filed a class action lawsuit in Los Angeles against Unocal as well as a Burmese government entity. The complaint alleged several human rights violations including torture, forced labor, crimes against humanity, violence against women, arbitrary arrest and detention, cruel, inhuman or degrading treatment, and wrongful death. Jurisdiction was premised primarily on the Alien Tort Statute. The lawsuit proceeded for 13 years, and there were numerous decisions by the district court and the Ninth Circuit Court of Appeals. On December 8, 2004, the parties notified the Ninth Circuit that they had reached a settlement. The case was subsequently dismissed when the settlement was finalized.[37] While the terms of the settlement were confidential, some reports suggest the settlement amount reached $30 million.[38]

In the 1990s, Holocaust survivors and their heirs filed numerous lawsuits in the United States and Europe seeking compensation for human rights violations committed during the Holocaust.[39] The defendants included multinational corporations (such as banks and insurance companies) as well as foreign governments. Claims ranged from the use of slave labor to summary execution.[40] Property claims involved stolen bank accounts, unpaid insurance policies, and confiscated real and personal property.[41] While most Holocaust-era lawsuits settled or were dismissed, this litigation played a significant role in the search for justice by Holocaust survivors and other victims of World War II-era atrocities.[42] As a result of these lawsuits, several foreign governments and multinational corporations entered negotiations with victims and survivor groups to address their claims. These negotiations produced significant financial settlements, which occurred, in part, because of the threat of continued litigation. As noted by the U.S. Supreme Court in *American Insurance Association v. Garamendi*, the willingness of defendants to settle these claims "was conditioned on some expectation of security from lawsuits in United States courts."[43]

While there has been a marked increase in the number of lawsuits alleging corporate liability for violations of international law, several recent decisions have raised questions regarding the applicability of transnational litigation to corporations. In *Presbyterian Church of Sudan v. Talisman*, for example, several Sudanese nationals filed a lawsuit against Talisman Energy for human rights violations arising out of Talisman's oil exploration projects in southern Sudan. The plaintiffs asserted Talisman's activities had led to forced displacement, extrajudicial killing, torture, rape and the destruction of villages, churches, and crops. After several rounds of litigation, the federal district court dismissed the lawsuit in 2006.[44] A key issue concerned whether Talisman had aided and abetted the human rights violations perpetrated by the Sudanese government. On appeal, the Second Circuit Court of Appeals affirmed the dismissal and imposed a more demanding standard on plaintiffs seeking to establish aiding and abetting liability. Specifically, it held that "a claimant must show that the defendant provided substantial assistance with the purpose of facilitating the alleged offenses."[45] Based on this standard, the Second Circuit dismissed the lawsuit against Talisman Energy, holding there was "no evidence that the company acted with the purpose of harming civilians living in southern Sudan."[46]

Another Second Circuit decision offers an even greater challenge to transnational litigation against corporations. In *Kiobel v. Royal Dutch Petroleum Co.*, Nigerian residents sued two oil exploration companies for their complicity in human rights abuses perpetrated by the Nigerian government. In 2006, the federal district court dismissed the plaintiffs' claims regarding aiding and abetting for property destruction, forced exile, extrajudicial killing, and violations of the rights to life, liberty, security, and association.[47] However, the district court allowed the claims for crimes against humanity, arbitrary arrest, torture, and other cruel, inhuman or degrading treatment to proceed. In 2010, the Second Circuit reversed the district court's finding and dismissed the case.[48] While acknowledging international law prohibited serious human rights abuses, the court found that international law did not recognize corporate liability for such actions.

> Because customary international law consists of only those norms that are specific, universal, and obligatory in the relations of States inter se, and because no corporation has ever been subject to any form of liability (whether civil or criminal) under the customary international law of human rights, we hold that corporate liability is not a discernable — much less universally recognized — norm of customary international law that we may apply pursuant to the ATS.[49]

The *Kiobel* approach to corporate liability under the Alien Tort Statute was not followed by other courts, including *Doe v. Exxon Mobil Corp.* and *Flomo v. Firestone Natural Rubber Co., LLC.*[50] In both these cases, the appellate courts held corporations could be sued under the Alien Tort Statute. Because of these competing positions, the U.S. Supreme Court granted certiorari in *Kiobel* on October 17, 2011.[51] The Court agreed to consider the following two questions:

1. Whether the issue of corporate civil tort liability under the Alien Tort Statute ("ATS"), 28 U.S.C. § 1350, is a merits question, as it has been treated by all courts prior to the decision below, or an issue of subject matter jurisdiction, as the court of appeals held for the first time.
2. Whether corporations are immune from tort liability for violations of the law of nations such as torture, extrajudicial executions or genocide, as the court of appeals decisions provides, or if corporations may be sued in the same manner as any other private party defendant under the ATS for such egregious violations, as the Eleventh Circuit has explicitly held.[52]

On February 28, 2012, the Supreme Court heard oral argument in *Kiobel* on the issue of corporate liability under the Alien Tort Statute.[53] In a surprising development, the Court subsequently announced it would rehear the case in Fall 2012 and asked the parties to file supplemental briefs on the following question: "Whether and under what circumstances the Alien Tort Statute, 28 U.S.C. § 1350, allows courts to recognize a cause of action for violations of the law of nations occurring within the territory of a sovereign other than the United States."[54] The Court's decision to consider whether the Alien Tort Statute could ever be used to recognize extraterritorial harms threatens the future of all ATS litigation. Oral argument is scheduled for Fall 2012 and a decision is expected in Spring 2013.

Transnational Litigation in the Amazon

Aguinda v. Texaco is one of the earliest cases to raise environmental claims from the Amazon in U.S. courts.[55] Seventy-six residents of the Oriente region of Ecuador filed the lawsuit in November 1993 in the federal district court for the Southern District of New York.[56] The lawsuit was filed as a class action on behalf of approximately 30,000 natives of the Amazon rainforest region. The lawsuit alleged Texaco polluted the rain forests and rivers of the Oriente region between 1964 and 1992 in violation of U.S. and international law. As a result of Texaco's operations, the plaintiffs alleged their health had been seriously harmed by increased cancer rates, genetic defects, and a host of other physical and mental injuries.[57] The complaint raised several causes of action, including negligence, public and private nuisance, strict liability, medical monitoring, trespass, civil conspiracy, and equitable relief. The complaint also raised the Alien Tort Statute as a separate cause of action. According to the complaint, the "Defendant's acts and omissions of intentionally and tortiously discharging crude oil and other toxins into the environment; in damaging the pristine rain forests of the Oriente; in destroying the streams, rivers, waterways and acquifers, and in threatening the survival of the indigenous people of the Oriente violate the law of nations, international law, worldwide industry standards and practices, as well as the laws of the United States."[58]

The issue of environmental harm under international law was addressed, albeit briefly, in the early stages of the litigation.[59] The district court noted that "[e]nvironmental damage is recognized in the domestic law of the United States as subject to legal restrictions."[60] The court also noted U.S. laws governing hazardous waste "may well prohibit the conduct alleged in the complaint if carried out in the United States."[61] With respect to international law, the court suggested it may also prohibit the misuse of hazardous waste, particularly when such action has transborder implications. In support of this argument, the district court referenced the Rio Declaration on Environment and Development, which provides that states have "the responsibility to ensure that activities within their jurisdiction or control do not cause damage to the environment of other States or areas beyond the limits of national jurisdiction."[62] While the district court indicated the lawsuit might proceed under the Alien Tort Statute "if there were established misuse of hazardous waste of sufficient magnitude to amount to a violation of international law," it found this decision must await additional information regarding actions taken in the United States.[63] The court added, however, that it should proceed with caution under the Alien Tort Statute if the underlying conduct occurred exclusively in a foreign country.

In December 1994, 23 Peruvian nationals (and four nonprofit organizations) filed a similar lawsuit, *Jota v. Texaco*, on behalf of approximately 25,000 Peruvian residents who lived downstream from the Oriente region.[64] This lawsuit was also based, in part, on the Alien Tort Statute. It also charged Texaco with numerous claims of environmental pollution. The plaintiffs sought monetary damages as well as extensive equitable relief to redress the damages that had been inflicted to the environment.

After several rounds of litigation, the district court consolidated the *Aguinda* and *Jota* cases and eventually dismissed them under the doctrine of *forum non conveniens* because the court believed it would be more convenient for the parties to litigate the case in Ecuador.[65] The doctrine of *forum non conveniens* involves a two-step process to determine whether a lawsuit may be dismissed in favor of a foreign forum.[66] First, the court must determine whether the foreign forum offers an adequate alternative forum. This is typically resolved if the defendant is amenable to process in the foreign forum. Second, the court must balance the public and private interest factors to determine whether dismissal is appropriate. The private interest factors include "the relative ease of access to sources of proof; availability of compulsory process for attendance of unwilling, and the cost of obtaining attendance of willing, witnesses; possibility of view of the premises, if view would be appropriate to the action; and all other practical problems that make trial of a case easy, expeditious and inexpensive."[67] The public interest factors include the administrative difficulties for courts in handling foreign cases, the burdens of imposing jury duty upon the people of a community with no relation to the litigation, the benefits of conducting a trial in the community where the harms occurred, and the problems associated with applying foreign law in domestic proceedings and the attendant problems in conflict of laws.[68] According to the district court, the two cases "have everything to do with Ecuador and nothing to do with the United States."[69] Dismissal was, therefore, warranted under the doctrine of *forum non conveniens*.

The Second Circuit affirmed this dismissal in August 2002.[70] First, the court held Ecuador offered an adequate alternate forum to hear the case.[71] The court indicated Ecuadorian courts could hear mass lawsuits alleging torts of this nature. The court also noted Ecuadorian courts would not be subject to corruption or undue influence. Second, the court found that public and private interest factors favored dismissal of the lawsuit.[72] The court indicated the litigation would be easier to conduct in Ecuador since the plaintiffs and the evidence were available there. In addition, the lack of a formal nexus between the plaintiffs' harms and the United States was also relevant. For these reasons, the Second Circuit affirmed the dismissal of the lawsuit on the condition that Texaco would agree to waive any possible statute of limitations defense that would apply between the original filing of the U.S. lawsuits and up to one year from the date of final judgment.[73]

As a result of the dismissal, a new lawsuit was filed in May 2003 with the Superior Court of Justice in Nueva Loja, Ecuador, which is located in the Oriente region. The lawsuit was filed against Chevron because Texaco had merged with Chevron in 2001. The lawsuit alleged Texaco and its subsidiary, TexPet, were liable for environmental contamination in the Oriente region. The lawsuit raised several demands:

(1) The elimination or removal of the contaminating elements that still threaten the environment and the health of the inhabitants.

a. The removal and the adequate treatment and disposition of the debris and contaminating materials still existing in the crude oil pits or ditches opened by Texaco and that have simply been covered up, and inadequately treated;
b. The cleaning of the rivers, estuaries, lakes, swamps and natural streams and artificial water streams, and the adequate disposition of all the debris material;
c. The removal of all the structural elements and machinery that are over the surface of the soil in the closed wells, stations, and Sub Stations, closed or abandoned, as well as the pipes (ducts), tubing, and other related elements along with the wells; and,
d. In general, the clearance of the terrains, plantations, crops, streets, roads and buildings where there may still exist contaminating residuals produced or generated as a consequence of the operations directed by Texaco, including the contaminating debris deposits built as a part of the wrongly [sic] environmental cleaning tasks.

(2) The repair of the environmental damages caused, in accordance to what is stated under Article 32 of the Law for Environmental Management.

a. That all the necessary works be done in the crude oil pits opened by Texaco, in order for them to recuperate the characteristics and natural conditions of the soil and of the adjacent terrains, that had prior to suffering the damages;
b. The hiring, at the Defendant's cost of people or specialized institutions in order for them to design and put into effect a recuperation plan of the fauna and flora;
c. The hiring at Defendant's cost of the people or specialized institutions in order for them to design and put into effect a plan to regenerate the aqueous life;
d. The hiring at Defendant's cost of the people or specialized institutions in order for them to design and put into effect a plan for improvement and monitoring to the health of the inhabitants of the affected towns (peoples) for the contamination.

The necessary recourses to cover the cost of the activities which are hereby

demanded, for the amount that it be determined by experts in accordance to the paragraph before last of Article 43 of the Law for Environmental Management, must be delivered to the Amazon Defence Front, in order for them to, with the concurrence of the advice from specialized International Institutions, it may apply exclusively, for the purposes determined in your verdict.
(3) The payment of a 10% of the total repair work, referred to in second paragraph of Article 43 of the law of environmental management; as well as the payment of the action costs and whatever is worth for the time and diligence employed in it, in accordance to Article 2261 of the Civil Code. Whatever it is ordered to be paid as per the foregoing concepts, it must be delivered to, by express request to that effect made by the Plaintiffs, to the Amazon Defence Front (FDA).[74]

This litigation proceeded for several years in Ecuadorian courts. Extensive evidence was compiled and reviewed by the courts. As the lawsuit progressed, Chevron increasingly argued that Ecuador was an inappropriate forum for the case and that the Ecuadorian courts were corrupt and incompetent. Chevron eventually filed its own action against the plaintiffs' lawyers in the case, alleging fraud and racketeering.[75]

On February 11, 2011, the Provincial Court of Justice issued a judgment in the case, holding that Chevron must pay $8.6 billion in damages: (1) $600 million for groundwater remediation; (2) $5.3 billion for soil remediation; (3) $200 million for damage to flora and fauna; (4) $150 million for drinking water remediation; (5) $1.4 billion for health care services; (6) $100 million for indigenous culture damages; and (7) $800 million for cancer deaths.[76] In addition, the judgment granted an additional $8.6 billion in punitive damages unless Chevron issued a public apology. Within a week of the court's ruling, the plaintiffs filed an appeal arguing that more money would be needed to redress the environmental harms.[77] Chevron denounced the ruling and vowed to appeal. According to a press release issued by Chevron, "[t]he Ecuadorian court's judgment is illegitimate and unenforceable. It is the product of fraud and is contrary to the legitimate scientific evidence. Chevron will appeal this decision in Ecuador and intends to see that justice prevails. United States and international tribunals already have taken steps to bar enforcement of the Ecuadorian ruling. Chevron does not believe that today's judgment is enforceable in any court that observes the rule of law. Chevron intends to see that the perpetrators of this fraud are held accountable for their misconduct."[78] In addition, Chevron sought a federal injunction to prevent the plaintiffs or their attorneys from enforcing the Ecuadorian judgment outside Ecuador. The injunction was issued by the federal district court for the Southern District of New York on March 7, 2011.[79] These proceedings remain ongoing.

Carijano v. Occidental Petroleum offers another example of transnational litigation involving the Amazon although the claims in this case are exclusively domestic in nature and not based on international norms. In 2007, several members of the Achuar indigenous group of Peru filed a lawsuit against Occidental Petroleum in California state court. Amazon Watch, Inc., a California-based environmental organization that works on behalf of indigenous peoples of the Amazon basin, also joined the lawsuit as a plaintiff. The lawsuit alleged that Occidental had caused extensive environmental harm in Peru through its oil extraction and production operations. According to the complaint:

> Oxy's oil operations in Lot 1AB contaminated Achuar land, and the waterways on which the Achuar people depend, with pollutants such as heavy metals, hydrocarbons, and other products of the oil industry. . . .
>
> Oxy separated crude oil at several separation batteries located along tributaries of the Corrientes River and the Macusari River, discharging millions of gallons of toxic produced waters into local waterways in the process. The produced waters contained harmful compounds including heavy metals such as lead, cadmium, chromium, mercury, and arsenic, organic compounds such as aromatic hydrocarbons, radioactive compounds, and cyanide, and were extremely saline. Oxy contaminated the local waterways with their release of these compounds. . . .
>
> Gases were also produced as part of the oil production and separation process, which Oxy burned through a flaring process. These flares release carbon dioxide, nitrogen oxide, and sulfur dioxide, which pollutes the air, contributes to climate change, and then returns to the soil and water sources in the form of acid rain.
>
> Oxy also stored chemical wastes and other toxic products in unlined earthen pits and pools and using other methods that failed to provide adequate protection against leaching and contaminating the environment. These toxic compounds also included heavy metals and organic compounds, many of which found their way into the local waterways. . . .
>
> Oxy also released oil and other compounds into the environment through frequent spills and ruptures of pipelines, which often sent crude oil directly into the Corrientes or Macusari River.[80]

The complaint raised several causes of action, including negligence, strict liability, battery, wrongful death, fraud and misrepresentation, public nuisance, private nuisance, and trespass. While the case was filed in California state court, it was eventually removed to federal court by the defendants.

On April 15, 2008, the district court dismissed the lawsuit, pursuant to the doctrine of *forum non conveniens*, finding that the case should be heard in Peru.[81] First, the court held that Peru offered an adequate alternate forum to hear the case.[82] The court found that Peruvian courts could hear cases involving civil liability of this nature, and that claims of corruption in the Peruvian courts were unsupported. Second, the court ruled that public and private interest factors favored dismissal of the lawsuit.[83] The court noted that travel costs and litigation costs, including translation services, also supported litigation in Peru. Finally, the court observed that a Peruvian tribunal would be better equipped to handle the case, which "raises issues such as the environmental regulation of Peruvian territory, and the allegedly tortuous conduct carried out against certain Peruvian citizens."[84]

On December 6, 2010, the Ninth Circuit Court of Appeals reversed the district court's dismissal.[85] The Ninth Circuit found the district court had abused its discretion on several aspects of the *forum non conveniens* analysis. In determining whether Peru was an adequate alternate forum, the court noted that Occidental had not indicated it was amenable to process in Peru if the case was dismissed. The failure of the district court to confirm Occidental's amenability to suit was an abuse of discretion because this would be the only way to ensure that Occidental would participate in the litigation.[86] The Ninth Circuit also determined that the district court abused its discretion by finding Peru could offer the plaintiffs a satisfactory remedy.[87] Significant evidence had been presented that Peruvian law would offer no practical remedy other than nominal damages to the plaintiffs. Significant evidence had also been presented that Peruvian courts were subject to discrimination and corruption. While the Ninth Circuit did not "hold that Peru is categorically an inadequate forum," the district court's analysis failed to consider the "troubling evidence of potential inadequacy."[88]

In considering the public and private interest factors, the Ninth Circuit again pointed to several limitations in the district court's analysis. It noted, for example, that the district court failed to give proper deference to the plaintiffs' selection of California as the proper forum for litigating the case.[89] This was particularly significant because Amazon Watch was a California plaintiff whose choice of California as the forum was entitled to a strong presumption. The district court also failed to consider whether a successful judgment against Occidental could be enforced in Peru.[90] There were questions regarding whether Occidental had available assets in Peru that could be used to satisfy the judgment. There were also questions regarding whether a U.S. court would enforce a Peruvian judgment.

On June 1, 2011, the Ninth Circuit granted a petition for rehearing and vacated its prior decision in *Carijano v. Occidental Petroleum*.[91] Its analysis differed slightly from the previously withdrawn opinion although the court once again reversed the district court's dismissal.

> Here the district court failed to consider all relevant private and public interest factors, entirely overlooking the enforceability of judgments factor, which weighs heavily against dismissal. It correctly assumed that Amazon Watch was a proper domestic plaintiff, but erroneously afforded reduced deference to its chosen forum and ignored the group entirely in the analysis of numerous factors. These errors led the district court to misconstrue factors that are neutral or weigh against dismissal, and to strike an unreasonable balance between the factors and the deference due a domestic plaintiff's chosen forum. Finally, the district court abused its discretion by failing to impose conditions on its dismissal that were warranted by facts in the record showing justifiable reasons to doubt Occidental's full cooperation in the foreign forum.[92]

These proceedings remain ongoing.

Aguinda and *Carijano* reveal the difficulties in pursuing transnational litigation for corporate abuses committed abroad. It is not surprising that the corporate defendants sought to dismiss these lawsuits under the doctrine of *forum non conveniens*. The doctrine requires courts to confront the reasons for litigating foreign harms in U.S. courts. While courts will afford deference to a plaintiff's choice of forum, they are mindful of the practical challenges associated with litigating these cases in the United States. In *Aguinda*, the Second Circuit accepted the argument that Ecuador was a more appropriate venue for the litigation. It is evident Chevron no longer takes this position. In *Carijano*, the Ninth Circuit rejected the argument that Peru was a more appropriate venue. While these lawsuits reached different outcomes, both *Aguinda* and *Carijano* reveal the difficulties in pursuing litigation in the country where the harms occurred and, therefore, they suggest the benefits of transnational litigation.[93]

Conclusion

Transnational litigation has been subject to numerous challenges. Some critics argue it represents a form of legal imperialism. Why should U.S. courts impose U.S. solutions on foreign litigants with foreign problems? Other critics assert it involves U.S. courts in delicate matters of foreign affairs. This explains why this litigation has been referred to as "plaintiff's diplomacy" or even "lawfare."[94] It is also asserted that the United States may not be the best forum for these cases: the events occurred abroad; witnesses and evidence may be located abroad. Finally, some critics allege these lawsuits hinder corporate operations abroad and that corporations are not liable for international law violations.[95]

There are equally compelling arguments in response to these criticisms. With respect to judicial imperialism, these cases generally do not involve the U.S. government as a litigant. Rather these cases are typically filed by foreign nationals, which undermines claims of imperialism. With respect to judicial interference in foreign affairs, courts have numerous rules and procedures that can address legitimate concerns. The political question doctrine or the act of state doctrine can limit cases that affect U.S. foreign policy.[96] The doctrine of *forum non conveniens* can remove cases from U.S. courts in appropriate circumstances. But, in many cases, the issue is not whether the United States is the best forum. Rather, the United States is often the only available forum. As *Carijano* reveals, some foreign jurisdictions may not be suitable for litigation. If a foreign forum is willing and able to hear the case, then it may be preferable to litigate the case in the forum where the injuries occurred. As *Aguinda* reveals, however, foreign litigation is fraught with its own complications.

Corporate liability is a contentious issue. It is certainly true that establishing liability for corporations may hinder their overseas operations. It is equally true, however, that corporations should not be immune from liability for harms committed abroad. To hold otherwise would embolden corporations to violate human rights and environmental law by transferring their operations to countries with limited regulatory schemes.

And, in fact, international law does regulate the actions of corporations. Some international norms explicitly address non-state actors such as corporations.[97] The Convention against Transnational Organized Crime, for example, requires each State Party "to establish the *liability of legal persons* for participation in serious crimes involving an organized criminal group . . . ."[98] Similarly, the Convention on Combating Bribery of Foreign Public Officials in International Business Transactions, requires each State Party "to establish the *liability of legal persons* for the bribery of a foreign public official."[99] There are several other treaties that also reference non-state actors.[100] Most international norms, however, do not distinguish between state and non-state actors. Since juridical persons are domestic creations, this is not surprising. Moreover, international law generally delegates enforcement of its norms to domestic mechanisms. Thus, the absence of international practice regarding corporate liability does not undermine the existence of such liability.[101] States must decide how to enforce international law against corporations within the context of their domestic legal systems.[102]

For these reasons, transnational litigation should be viewed as an appropriate mechanism for norm enforcement when corporations violate international law in their overseas operations. Such litigation can play a critical role in addressing environmental harms and human rights violations throughout the world. Indeed, transnational litigation has brought to light many of the environmental harms perpetrated in the Amazon basin and forced corporations to respond in court. If the U.S. Supreme Court strikes down the extraterritorial reach of the ATS in the *Kiobel* case, it will have implications throughout the world and will force victims from the Amazon basin to continue the search for justice.

Acknowledgments

The author would like to thank Yesenia Acosta, Regina Bagdasarian, Christopher Baidoo, Brian Israel, and Kirstin Jarstad for their research assistance. He also appreciates the comments and support of Professor James Cooper. Portions of this chapter are based on WILLIAM J. ACEVES, THE ANATOMY OF TORTURE: A DOCUMENTARY HISTORY OF *FILARTIGA V. PENA-IRALA* (2007). Professor Aceves participated in several lawsuits described in this chapter.

Notes

1 *See generally* STEPHEN MCCAFFERY & THOMAS MAIN, TRANSNATIONAL LITIGATION IN COMPARATIVE PERSPECTIVE (2009); Harold HONGJU KOH, TRANSNATIONAL LITIGATION IN UNITED STATES COURTS (2008).

2 *See generally* Beth Stephens, *Translating Filartiga: A Comparative and International Law Analysis of Domestic Remedies for International Human Rights Violations*, 27 YALE J. INT'L L. 1 (2002); William J. Aceves, *Liberalism and International Legal Scholarship: The Pinochet Case and the Move Toward a Universal System of Transnational Law Litigation*, 41 HARV. INT'L L. J. 129 (2000); Harold Hongju Koh, *Transnational Legal Process*, 75 NEB. L. REV. 181 (1996); Harold Hongju Koh, *Transnational Public Law Litigation*, 100 YALE L.J. 2347 (1991).

3 *See generally* MICHAEL KOEBELE, CORPORATE RESPONSIBILITY UNDER THE ALIEN TORT STATUTE (2009).

4 *Id.* at 10. *See also* SARAH JOSEPH, CORPORATIONS AND TRANSNATIONAL HUMAN RIGHTS LITIGATION (2004); MENNO T. KAMMINGA & SAMAN ZIA-ZARIFI, LIABILITY OF MULTINATIONAL CORPORATIONS UNDER INTERNATIONAL LAW (2001).

5 *See, e.g.*, David Scheffer & Caroline Kaeb, *The Five Levels of CSR Compliance: Resiliency of Corporate Liability under the Alien Tort Statute and the Case for a Counterattack Strategy in Compliance Theory*, 29 BERKELEY J. INT'L L. 334 (2011); Ronald C. Slye, *Corporations, Veils, and International Criminal Liability*, 33 BROOKLYN J. INT'L L. 955 (2008); Sarah Cleveland, *The Alien Tort Statute, Civil Society, and Corporate Responsibility*, 56 RUTGERS L. REV. 971 (2005); Beth Stephens, *Corporate Liability Before and After* Sosa v. Alvarez-Machain, 56 RUTGERS L. REV. 995 (2004); Steven R. Ratner, *Corporations and Human Rights: A Theory of Legal Responsibility*, 111 YALE L.J. 443 (2001).

6 *See, e.g.*, Li-Wen Lin, *Corporate Social Responsibility in China: Window Dressing or Structural Change?*, 28 BERKELEY J. INT'L L. 64 (2010); Lauren A. Dellinger, *Corporate Social Responsibility: A Multifaceted Tool to Avoid Alien Tort Claims Act Litigation While Simultaneously Building a Better Business Reputation*, 40 CAL. W. INT'L L.J. 55 (2009); Christopher McCrudden, *Human Rights Codes for Transnational Corporations: What Can the Sullivan and MacBride Principles Tell Us?*, 19 OXFORD J. LEGAL STUD. 167 (1999); Johanna Elisabeth Maria Kolk et al., *International Codes of Conduct and Corporate Social Responsibility: Can Transnational Corporations Regulate Themselves?*, 8 TRANSNAT'L CORPS. 143 (1999).

7 For a comparative approach to transnational litigation, see TORTURE AS TORT: COMPARATIVE PERSPECTIVES ON THE DEVELOPMENT OF TRANSNATIONAL HUMAN RIGHTS LITIGATION (Craig Scott ed., 2001).

8 *See generally* Theodore Macdonald, *Amazonian Indigenous Views on the State: A Place for Corporate Social Responsibility*, 33 SUFFOLK TRANSNAT'L L. REV. 439 (2010); Megan S. Chapman, *Seeking Justice in Lago Agrio and Beyond: An Argument for Joint Responsibility of Host States and Foreign Investors Before the Regional Human Rights Systems*, HUM. RTS. BRIEF 6 (2010); Sarah M. Morris, *The Intersection of Equal and Environmental Protection: A New Direction for Environmental Alien Tort Claims After* Sarei *and* Sosa, 41 COLUM. HUM. RTS. L. REV. 275 (2009); Judith Kimerling, *Transnational Operations, Bi-National Injustice: ChevronTexaco and Indigenous Huaorani and Kichwa in the Amazon Rainforest in Ecuador*, 31 AM. INDIAN L. REV. 445 (2006/07); Judith Kimerling, *Indigenous Peoples and the Oil Frontier in Amazonia: The Case of Ecuador, Chevron-Texaco, and* Aguinda v. Texaco, 38 N.Y.U. J. INT'L L. & POL. 413 (2006).

9 Some lawsuits focus solely on human rights abuses, unrelated to environmental harms. In *Sinaltrainal v. Coca-Cola*, for example, members of a Colombian trade union filed a federal lawsuit in the United States against Coca-Cola, alleging its complicity in human rights abuses in Colombia. *See* Sinaltrainal v. Coca-Cola Company, 578 F.3d 1252 (11th Cir. 2009). Other lawsuits address corporate activity in the extraction industry although the underlying complaints do not raise environmental claims. In *Mujica v. Occidental Petroleum Corp.*, for example, several Colombian nationals filed a lawsuit against Occidental Petroleum Inc. and AirScan Inc., charging them with complicity in the 1998 bombing of civilians in the village of Santo Domingo, Colombia. The plaintiffs alleged several violations of international law, including war crimes, crimes against humanity, extrajudicial killing, torture, and cruel, inhuman, and degrading treatment. *See* Mujica v. Occidental Petroleum Corp., 564 F.3d 1190 (9th Cir. 2009). This chapter focuses on lawsuits involving environmental harms.

10 Aguinda v. Texaco, Inc., 303 F.3d 470 (2d Cir. 2002).

11 Carijano v. Occidental Petroleum Corp. 626 F.3d 1137 (9th Cir. 2010).

12 *See, e.g.*, Dennick v. Railroad Company, 103 U.S. 11 (1880); McKenna v. Fisk, 42 U.S. 241 (1843).

13 BETH STEPHENS ET AL., INTERNATIONAL HUMAN RIGHTS LITIGATION IN U.S. COURTS 43 (2d ed. 2008).

14 *See* GARY BORN & PETER RUTLEDGE, INTERNATIONAL CIVIL LITIGATION IN UNITED STATES COURTS (4th ed. 2007).

15 Morrison v. Nat'l Australia Bank, 130 S. Ct. 2869 (2010).

16 Ralph G. Steinhardt, *International Humanitarian Law in the United States:* Yamashita, Filartiga, *and 9/11*, 36 GEO. WASH. L. REV. 1, 22 (2004).

17 *See generally* PETER HENNER, HUMAN RIGHTS AND THE ALIEN TORT STATUTE (2009); JEFFREY DAVIS, JUSTICE ACROSS BORDERS: THE STRUGGLE FOR HUMAN RIGHTS IN U.S. COURTS (2008).

18 28 U.S.C. § 1350.

19 *See generally* Jordan Paust, *The History, Nature, and Reach of the Alien Tort Claims Act*, 16 FLA. J. INT'L L. 249 (2004); William S. Dodge, *The Historical Origins of the Alien Tort Statute: A Response to the "Originalists,"* 19 HASTINGS INT'L & COMP. L. REV. 221 (1996); Joseph Modeste Sweeney, *A Tort Only in Violation of the Law of Nations*, 18 HASTINGS INT'L & COMP. L. REV. 445 (1995); Anne-Marie Burley, *The Alien Tort Statute and the Judiciary Act of 1789: A Badge of Honor*, 83 AM. J. INT'L L. 461 (1989).

20 *See generally* WILLIAM J. ACEVES, THE ANATOMY OF TORTURE: A DOCUMENTARY HISTORY OF *FILARTIGA V. PENA-IRALA* (2007). *See also* Jeffrey M. Blum & Ralph G. Steinhardt, *Federal Jurisdiction over International Human Rights Claims: The Alien Tort Claims Act After* Filartiga v. Pena-Irala, 22 HARV. INT'L L.J. 53 (1981).

21 Filartiga v. Pena-Irala, No. 79 C 917, slip op. (E.D.N.Y. May 15, 1979).

22 Filartiga v. Pena-Irala, 630 F.2d 876 (2d Cir. 1980).

23 *Id.* at 878.

24 *Id.* at 890.
25 ACEVES, *supra* note 20, at 8.
26 STEPHENS ET AL., *supra* note 13 at 139–205; HENNER, *supra*, at 133–139.
27 STEPHENS ET AL., *supra* 13 at 281–333; HENNER, *supra*, at 207–242.
28 Sosa v. Alvarez-Machain, 542 U.S. 692 (2004).
29 *Id.* at 725.
30 The *Sosa* decision has been subject to extensive commentary. *See generally* Ehren J. Brav, *Recent Development: Opening the Courtroom Doors to Non-Citizens: Cautiously Affirming Filartiga for the Alien Tort Statute*, 46 HARV. INT'L L.J. 265 (2005); Harold Koh, *The Supreme Court Meets International Law*, 12 TULSA J. COMP. & INT'L L. 1 (2004); Gary Clyde Hufbauer, *What's the Sequel to* Sosa v. Alvarez-Machain*?*, 12 TULSA J. COMP. & INT'L L. 77 (2004); Note, *The Offenses Clause After* Sosa v. Alvarez-Machain, 118 HARV. L. REV. 2378 (2005); Ralph G. Steinhardt, *Laying One Bankrupt Critique to Rest:* Sosa v. Alvarez-Machain *and the Future of International Human Rights Litigation in U.S. Courts*, 57 VAND. L. REV. 2241 (2004); Beth Stephens, Sosa v. Alvarez-Machain*: "The Door Is Still Ajar" For Human Rights Litigation in U.S. Courts*, 70 BROOK. L. REV. 533 (2004); Eugene Kontorovich, *Implementing* Sosa v. Alvarez-Machain*: What Piracy Reveals About the Alien Tort Statute*, 80 NOTRE DAME L. REV. 111 (2004).
31 *See generally* Philip Mariani, *Assessing the Proper Relationship Between the Alien Tort Statute and the Torture Victim Protection Act*, 156 U. PA. L. REV. 1383 (2008); Eric Engle, *The Alien Tort Statute and the Torture Victims' Protection Act: Jurisdictional Foundations and Procedural Obstacles*, 14 WILLAMETTE J. INT'L L. & DISPUTE RES. 1 (2006); Christopher W. Haffke, *Torture Victim Protection Act: More Symbol Than Substance*, 43 EMORY L.J. 1467 (1994); Rachel Schwartz, *"And Tomorrow?" The Torture Victim Protection Act*, 11 ARIZ. J. INT'L & COMP. L. 271 (1994); Yoav Gery, *The Torture Victim Protection Act: Raising Issues of Legitimacy*, 26 GEO. WASH. J. INT'L L. & ECON. 597 (1993).
32 28 U.S.C. § 1350 (note).
33 *Id.* at § 2(a).
34 *Id.* at §§ 2(b) and (c).
35 18 U.S.C. § 1331; 28 U.S.C. § 1605.
36 *See generally* Robert C. Thompson et al., *Translating Unocal: The Expanding Web of Liability for Business Entities Implicated in International Crimes*, 40 GEO. WASH. INT'L L. REV. 841 (2009); Richard L. Herz, Doe v. Unocal *and the Poverty of the Corporate Attack on the Alien Tort Statute*, 56 RUTGERS L. REV. 1005 (2004); Terry Collingsworth, *The Key Human Rights Challenge: Developing Enforcement Mechanisms*, 15 HARV. HUM. RTS. J. 183 (2002).
37 Doe I v. Unocal Corp., 403 F.3d 708 (9th Cir. 2005). *See generally* William J. Aceves, *International Decisions: Doe v. Unocal*, 92 AM. J. INT'L L. 309 (1998).
38 Paul Magnusson, *A Milestone for Human Rights*, BUS. WK., Jan. 24, 2005, at 63; Marc Lifsher, *Unocal Settles Human Rights Lawsuit Over Alleged Abuses at Myanmar Pipeline*, LOS ANGELES TIMES, Mar. 22, 2005, at C1.
39 *See generally* MICHAEL BAZYLER, HOLOCAUST JUSTICE: THE BATTLE FOR RESTITUTION IN AMERICA'S COURTS (2003); STUART E. EIZENSTAT, IMPERFECT JUSTICE: LOOTED ASSETS, SLAVE LABOR, AND THE UNFINISHED BUSINESS OF WORLD WAR II (2003); JANE SCHAPIRO, INSIDE A CLASS ACTION: THE HOLOCAUST AND THE SWISS BANKS (2003).
40 *See, e.g., In re* Holocaust Victim Assets Litig., 1998 U.S. Dist. LEXIS 18014 (E.D.N.Y. Oct. 7, 1998) (Swiss banks); Burger-Fischer v. Degussa AG, 65 F. Supp. 2d 248 (D.N.J. 1999) (German corporations); *In re* Austrian & German Bank Holocaust Litig., 80 F. Supp. 2d 164 (S.D.N.Y. 2000) (German and Austrian banks); *In re* Assicurazioni Generali S.P.A. Holocaust Ins. Litig., Judicial Panel on Multi Dist. Litig., 2000 U.S. Dist. LEXIS 17853 (insurance companies); Bodner v. Banque Paribas, 114 F. Supp. 2d 117 (E.D.N.Y.

2000) (French banks); Abrams v. Société Nationale Des Chemins De Fer Francais, 389 F.3d 61 (2d Cir. 2004) (French railroad).

41 Lawsuits were also filed by other victims of human rights abuses committed during World War II, including U.S. prisoners of war as well as Chinese and Korean women who were victims of sexual violence. *See, e.g.* Hwang Geum Joo v. Japan, 332 F.3d 679 (D.C. Cir. 2003); *In re* World War II Era Japanese Forced Labor Litig., 164 F. Supp. 2d 1153 (N.D. Cal. 2001).

42 *See* Burt Neuborne, *Preliminary Reflections on Aspects of Holocaust-era Litigation in American Courts*, 80 WASH. U. L.Q. 795 (2002); Morris A. Ratner, *The Settlement of Nazi-era Litigation Through the Executive and Judicial Branches*, 20 BERKELEY J. INT'L L. 212 (2002).

43 Am. Ins. Ass'n v. Garamendi, 539 U.S. 396, 405 (2003).

44 Presbyterian Church of Sudan v. Talisman Energy, Inc., 453 F. Supp. 2d 633 (S.D.N.Y. 2006).

45 Presbyterian Church of Sudan v. Talisman Energy, Inc., 582 F.3d 244 (2d Cir. 2009).

46 *Id.* at 248.

47 Kiobel v. Royal Dutch Petroleum Co., 456 F. Supp. 2d 457 (S.D.N.Y. 2006).

48 Kiobel v. Royal Dutch Petroleum Co., 621 F.3d 111 (2d Cir. 2010).

49 *Id.* at 149.

50 Doe v. Exxon Mobil Corp., 654 F.3d 11 (D.C. Cir. 2011); Flomo v. Firestone Natural Rubber Co., LLC, 643 F.3d 1013 (7th Cir. 2011).

51 The U.S. Supreme Court granted certiorari in a second case, *Mohamad v. Rajoub*, 634 F.3d 604 (D.C. Cir. 2011), to address whether the Torture Victim Protection Act applies to non-state actors. On April 18, 2012, the Supreme Court held in *Mohamad* determined that corporations could not be sued under the TVPA.

52 Questions Presented, Kiobel v. Royal Dutch Petroleum Co., available at http://www.supremecourt.gov/qp/10-01491qp.pdf.

53 The *Kiobel* decision has already received extensive commentary. *See generally* Frank Cruz-Alvarez & Laura E. Wade, *The Second Circuit Correctly Interprets the Alien Tort Statute:* Kiobel v. Royal Dutch, U. MIAMI L. REV. 1109 (2011); Geoffrey Pariza, *Genocide, Inc.: Corporate Immunity to Violations of International Law After* Kiobel v. Royal Dutch Petroleum, 8 LOY. U. CHI. INT'L L. REV. 229 (2011); Janine M. Stanisz, *The Expansion of Limited Liability Protection in the Corporate Form: The Aftermath of* Kiobel v. Royal Dutch Petroleum Co., 5 BROOK. J. CORP. FIN. & COM. L. 573 (2011).

54 Questions Presented, Kiobel v. Royal Dutch Petroleum Co., available at http://www.supremecourt.gov/qp/10-01491qp.pdf

55 *See generally* Lawrence Hurley, *The Tort Heard Round the World*, CAL. LAWYER, Feb. 2012, at 12.

56 Complaint, Aguinda v. Texaco, Inc., 1994 U.S. Dist. LEXIS 4718 (S.D.N.Y. 1994).

57 In 2001, Texaco merged into Chevron. As a result, Chevron became responsible for defending the lawsuit based on Texaco's actions in Ecuador.

58 Complaint, Aguinda v. Texaco, *supra* note 56, at 35.

59 Aguinda v. Texaco, Inc., 1994 U.S. Dist. LEXIS 4718 (S.D.N.Y. 1994).

60 *Id.* at *23.

61 *Id.* at *24.

62 *Id.* at *22 (quoting Principle 2 of the Rio Declaration on Environment and Development).

63 *Id.* at *24.

64 *See generally* Joanna E. Arlow, *The Utility of ATCA and the "Law of Nations" in Environmental Torts Litigation:* Jota v. Texaco, Inc. *and Large Scale Environmental Destruction*, 7 WIS. ENVT'L. L.J. 93 (2000); Peggy Rodgers Kalas, *The Implications of* Jota v. Texaco *and the Accountability of Transnational Corporations*, 12 PACE INT'L L. REV. 47 (2000).

65 Aguinda v. Texaco, Inc., 142 F. Supp. 2d 534 (S.D.N.Y. 2001).
66 *See generally* Piper Aircraft Co. v. Reyno, 454 U.S. 235, 255 (1981).
67 Gulf Oil Corp. v. Gilbert, 330 U.S. 501, 508 (1947).
68 *Id.* at 508–509.
69 Aguinda v. Texaco, 142 F. Supp. 2d at 537.
70 Aguinda v. Texaco, Inc., 303 F.3d 470 (2d Cir. 2002).
71 *Id.* at 476–479.
72 *Id.* at 479–480.
73 *Id.* at 480.
74 Complaint at 22–25, Lawsuit for Alleged Damages Filed Before the President of the Superior Court of "Nueva Loja" in Lago Agrio, Province of Sucumbios; on May 7, 2003, by 48 Inhabitants of the Orellana and the Sucumbios Province (Sup. Ct. Nueva Loja May 7, 2003), available at http://chevrontoxico.com/assets/docs/2003-ecuador-legal-complaint.pdf.
75 *See Chevron Files Fraud and RICO Case Against Lawyers and Consultants Behind Ecuador Litigation*, ENP Newswire, Feb. 4, 2011.
76 Chevron Corp. v. Donziger, 11 Civ. 0691, slip op. at 53 (S.D.N.Y. Mar. 7, 2011). *See generally* Simon Romero & Clifford Krauss, *Chevron Is Ordered to Pay $9 Billion by Ecuador Judge*, New York Times, Feb. 15, 2011, at A4.
77 *Ecuador: Appeals Filed in Chevron Case*, New York Times, Feb. 18, 2011, at A6.
78 Press Release, Illegitimate Judgment against Chevron in Ecuador Lawsuit, (Feb. 14, 2011), available at http://www.chevron.com.
79 Chevron Corp. v. Donziger, 11 Civ. 0691, slip op. (March 7, 2011).
80 Complaint at 6–7, Carijano v. Occidental Petroleum Corp., 548 F. Supp. 2d 823 (C.D. Cal. 2008).
81 Carijano v. Occidental Petroleum Corp., 548 F. Supp. 2d 823 (C.D. Cal. 2008).
82 *Id.* at 832.
83 *Id.* at 832–834.
84 *Id.* at 834.
85 Carijano v. Occidental Petroleum, 626 F.3d 1137 (9th Cir. 2010).
86 *Id.* at 1145.
87 *Id.* at 1146.
88 *Id.* at 1148.
89 *Id.* at 1148–1151.
90 *Id.* at 1153.
91 Carijano v. Occidental Petroleum, 643 F.3d 1216 (9th Cir. 2011).
92 *Id.* at 1236.
93 *See generally* Cassandra Burke Robertson, *Transnational Litigation and Institutional Choice*, 51 B.C. L. Rev. 1081 (2010); M. Ryan Casey & Barrett Ristroph, *Boomerang Litigation: How Convenient is Forum Non Conveniens?*, 4 BYU Int'l L. & Mgmt. Rev.21 (2007).
94 *See, e.g.*, David B. Rivkin, Jr. & Lee Casey, *Crimes Outside the World's Jurisdiction*, New York Times, July 22, 2003, at A19; Anne-Marie Slaughter & David Bosco, *Plaintiff's Diplomacy*, Foreign Aff., Sept.–Oct. 2000, at 102.
95 Transnational litigation against corporations has been subject to extensive criticism. *See, e.g.*, Julian G. Ku, *The Curious Case of Corporate Liability Under the Alien Tort Statute: A Flawed System of Judicial Lawmaking*, 51 Va. J. Int'l L. 353 (2011); Jonathan Drimmer, *Think Globally, Sue Locally: Out-of-Court Tactics Employed by Plaintiffs, Their Lawyers, and Their Advocates in Transnational Tort Cases* (U.S. Chamber Institute for Legal Reform, Washington, D.C.), June 2010.
96 *See Baker v. Carr*, 369 U.S. U.S. 186 (1962) (discussing the political question doctrine). *See W.S. Kirkpatrick & Co., Inc. v. Envtl. Tectonics Corp., Int'l*, 493 U.S. 400 (1990) (discussing the act of state doctrine).

97 *See generally* Jordan Paust, *Nonstate Actor Participation in International Law and the Pretense of Exclusion*, 51 Va. J. Int'l L. 977 (2011); Harold Hongju Koh, *Separating Myth from Reality About Corporate Responsibility Litigation*, 7 J. Int'l Econ. L. 263, 265 (2004).

98 Convention against Transnational Organized Crime, Nov. 15, 2000, art. 10(1), 2225 U.N.T.S. 209 (emphasis added).

99 Convention on Combating Bribery of Foreign Public Officials in International Business Transactions, Dec. 17, 1997, art. 2, S. Treaty Doc. No. 105–43 (emphasis added).

100 *See, e.g.*, European Convention on the Prevention of Terrorism, May 16, 2005, art. 10(1), C.E.T.S. No. 196 (2005) ("Each Party shall adopt such measures as may be necessary, in accordance with its legal principles, to establish the *liability of legal entities* for participation in the offences set forth in Articles 5 to 7 and 9 of this Convention.") (emphasis added); Basel Convention on the Control of Transboundary Movements of Hazardous Wastes and Their Disposal, Mar. 22, 1989, 1673 U.N.T.S. 57; (establishing liability of legal persons for harms incurred during transboundary movement of hazardous waste); International Convention on the Suppression and Punishment of the Crime of Apartheid, Nov. 3, 1973 art. I(2), 1015 U.N.T.S. 243 ("The States Parties to the present Convention declare criminal those *organizations, institutions, and individuals* committing the crime of apartheid.") (emphasis added); International Convention on Civil Liability for Oil Pollution Damage, Nov. 29, 1969, 973 U.N.T.S. 3 (establishing liability of ship owners for oil pollution and defining owner to include a company).

101 Indeed, the existence of individual criminal responsibility for corporate officials, as evidenced in numerous Nuremberg and post-Nuremberg proceedings, does not undermine the existence of corporate liability. *See generally* Wolfgang Kaleck & Miriam Saage-Maab, *Corporate Accountability for Human Rights Violations Amounting to International Crimes*, 8 J. INT'L CRIM. JUST. 699 (2010).

102 *See* Eileen Denza, *The Relationship Between International and National Law, in* INTERNATIONAL LAW 411 (Malcolm D. Evans et al., 3d ed. 2010); 1 ROBERT JENNINGS & ARTHUR WATTS, OPPENHEIM'S INTERNATIONAL LAW § 21 (9th ed. 1992).

4 The Yasuní Ishpingo-Tambococha-Tiputini Initiative: Creative Problem Solving in Amazonia

James Cooper

On August 3, 2010, the United Nations Development Programme (UNDP) and the government of Ecuador agreed to establish a trust fund to protect the country's Yasuní National Park Ishpingo-Tambococha-Tiputini (ITT) oil fields from drilling ("the Yasuní Initiative"). The scheme is designed to prevent discharge into the earth's atmosphere of more than 400 million metric tons of carbon estimated to result if the petroleum were to be extracted and burned as fuel. The Yasuní Initiative would save the local rain forest from the ravages of oil development that has stained other parts of Ecuador.[1] The fund is to hold some 50 percent of the estimated US$7 billion that would have been earned had the petroleum been drilled.

This contribution builds on Richard Finkmoore's and William Aceves' chapters examining international legal regimes and how they apply to the Amazon basin — particularly those that have come out of the United Nations, with an emphasis on a specific mechanism that has been designed to save the rain forests in the Amazon: The Yasuní Initiative, first proposed by Ecuador's President Rafael Correa at the United Nations' climate change conference on September 24, 2007. The Yasuní Initiative offers a novel approach to carbon emissions control mixed with sustainable development. The government of Ecuador is to sell carbon emission credits to European and other industrialized powers in exchange for not drilling for an estimated 870 million barrels of heavy crude oil in Yasuní National Park, a remote 2.4 million acre rain forest, located 190 miles from the country's capital Quito and alongside the border with neighboring Perú. The plan also seeks international donations of $3 billion over the next decade to keep this oil — 20 percent of the country's known reserves — under the ground.

The Yasuní Initiative is an example of the kind of out of the box thinking, sometimes called Creative Problem Solving, that provides novel approaches to solving complex problems. The late Janeen Kerper and I have defined Creative Problem Solving as "an approach to law and lawyering that views legal problems beyond the litigation paradigm . . . a creative problem solver understands the nuances of group process, cross-cultural communications and interdisciplinary collaboration in solving a problem."[2] It has been applied in a number of domestic contexts. For example, it has provided a framework for preventive lawyering;[3] it has championed judicial innovations like drug treatment court,[4] teen court, and alternative sentencing;[5] and it has provided the underpinnings for more holistic resolutions of disputes.[6]

There have also been international and transnational applications of Creative Problem Solving. It has been the umbrella term used to describe new approaches to resolve long-standing international disputes including that between Cuba and the United States.[7] Rhona K.M. Smith has applied Creative Problem Solving techniques when addressing problems encountered by indigenous peoples.[8] Dino Kritsiotis has examined the manner in which international law has been used as a form of problem solving.[9] Negotiation has long been viewed as a problem-solving tool in the international arena.[10] Richard Falk reflected on Tom Farer's analysis of the problem solving capacity of the international community.[11] Clearly, the Yasuní Initiative is an example of problem-solving. The plan engages a number of stakeholders — the local community of indigenous people, the peoples of the Amazon, the donor agencies and countries, and, because it will assist in reducing carbon emissions, the globe.

The Yasuní Initiative is also fundamentally an integrative, preventive approach to the growing problem of carbon emission[12] and the chronic underdevelopment of the Amazon basin is interdisciplinary in nature as it directly affects indigenous self-determination, the international right to development, and the very principles of state sovereignty. The Ecuadorian government seeks to prevent environmental damage to the Amazon region and the negative effects to its residents, which has occurred in the past with previous oil exploration projects.

The forward-looking Yasuní Initiative contrasts with traditional legal methods. In this book, William Aceves discusses the Ecuadorian case against Texaco/Chevron using the Alien Torts Claims Act in U.S. courts,[13] a classic backwards-looking approach of "Who did what to whom and who is responsible for it?" Oil spills around Ecuador's Lago Agrio have created health issues and blighted the environment, according to a $27-billion lawsuit filed against Chevron by Ecuadorian environmentalists.[14] Chevron denies that it is at fault. The company, in 2001, acquired Texaco, which explored for and extracted oil in Lago Agrio beginning in 1964 in partnership with Ecuador's state oil company, Petroecuador.

The first part of this chapter examines the context for the Yasuní Initiative and its antecedents, which at first blush appear to be a classic dilemma of economic development pressures versus environmental preservation. Oil is Ecuador's main export, financing roughly 35 percent of Ecuador's public spending. For years the country has produced an average of about 500,000 barrels per day of crude, of which state-owned Petroecuador accounts for just over 50 percent. By 2011, however, the daily production level has fallen to 300,000 barrels. Oil revenue is critical to the Ecuadorian economy.[15] Though half of Ecuador lies in the Amazon basin, its rainforest is shrinking faster than in neighboring countries (by 1.67% a year).[16] Clearly, there is much incentive to exploit further oilfields to provide for the social needs of Ecuador, a developing country rife with major challenges.

The second part of this chapter explores these needs, in particular those of the indigenous communities most affected by extractive activities in the Yasuní National Park. The Yasuní Initiative is forward-looking toward sustainable development: "The price mechanism enables compensation for continued avoidance of carbon dioxide emissions which will thereby safeguard both the habitat of Yasuní and the people who have lived sustainably in that habitat."[17]

The third part of this chapter deals with some of the challenges that President Correa faces as he pursues the Yasuní Initiative. First, it is not easy to enroll international sponsors -sovereign wealth funds and philanthropic entities — to finance the highly complex problem of determining who should pay for the preservation of the rain forest. The second major challenge is that constitutional issues may arise as the scheme entails domestic agreements with the Ecuadorian state and other interested parties. President Rafael Correa's administration must comply with the activist, indigenous–empowering constitution that was promulgated in Ecuador in 2008. Indeed, it is unclear whether a simple presidential decree is sufficient constitutionally or whether a more public process is required to implement a new paradigm through which to build value and protect the common public good. How much power can the purchasers of these lands exercise over sovereign countries' development agendas? What role do transparency and the rule of law play in Ecuador? The lack of constitutional legal clarity about the Ecuadorian Constitution must give pause. That many foreign donors are pulling out of the Yasuní Initiative does not bode well for its future.[18]

This chapter then concludes with some thoughts about whether the Yasuní Initiative can be a model for future sustainable development schemes in the Amazon, and where the environment and marginalized populations require protection. It could very well set a precedent in the fight against global climate change and lower the high cost of going green. That is, if the Yasuní Initiative ever gets funded and the difficulties concerning the interpretation of the 2008 Ecuadorian Constitution's provisions on natural resources are ever resolved. But first, this chapter examines the history of the Yasuní Initiative before we deal with the challenges concerning the project coming to fruition.

The History Behind Ecuador's Yasuní Initiative

In 1979, Ecuador designated the Yasuní Forest a national park, deeming it a natural reserve worthy of preservation. In 1989, UNESCO designated the Yasuní National Park a Biosphere Reserve.[19] The park is special for a number of reasons: It is one of the most bio-diverse places on the planet: "The highest observed species richness of trees in small plots has been recorded in the western Amazonia."[20]

Within its 2.4 million acres of pristine forests, there are some 4,000 species of plants, 173 species of animals, and over 100,000 species of insects per hectare. The Yasuní National Park is also home to indigenous tribes — the Huaroani, the Tagaeri, and the Taromenane — who all hunt and gather in its remote interior, living traditional lifestyles. There are an estimated 9,800 people living within the park. It is also home to a unique breed of small bat. The harpy eagle and jaguar are also found in the Yasuní National Park. These most rare of animals require large expansive areas, and not small reserves, in order to hunt and thrive.

The Yasuní Initiative calls for oil reserves to remain underground in exchange for "fair compensation" from the international community for 50 percent of the lost revenues from choosing not to drill. The money is to be used for renewable energy projects and other environmental activities. With oil hovering above $100 per barrel (2011), more than the price when Ecuador's President first made his proposal public in 2007, there are estimates of $9.2 billion in revenue that would be realized by the oil extraction. President Correa, who earned a Ph.D. degree in Economics from the University of Illinois at Urbana-Champaign, estimates that the international community should contribute $350 million annually for a decade, which would be the revenue that Ecuador would earn if the oil would be extracted.

In addition to the money it would earn from this arrangement, Ecuador could also sell greenhouse gas emission credits that would result from foregoing development of the ITT reserve. Under the Kyoto Protocol of December 1997, greenhouse gas emission targets were not mandatory for developing countries. The targets, however, are binding for industrialized countries. But the Protocol also allows various mechanisms for international emissions trading. Through the "Clean Development Mechanism," industrialized countries may invest in projects that reduce emissions in developing countries and credit a portion of these reductions to their own emissions reduction targets. Such a carbon market attempts "to rectify the effects of missing property rights on resources within poorer nations by creating global rights on the use of resources by wealthy nations."[21] European countries could accept the Yasuní certificates as an equivalent to the carbon allowances traded on the European Energy Exchange in Leipzig, Germany. The exchange is a market for energy products such as electricity and natural gas, as well as carbon dioxide emission rights.

The money from selling carbon-emission certificates would go to a trust fund managed by international organizations. The interest would be spent on protecting national parks and funding alternative energy projects and other environmental initiatives. The local communities would work on projects that do not damage the environment yet produce some economic benefit while still preserving cultural practices.

Ecuador Needs the Funds from the Yasuní Initiative

The funds to be generated from the Yasuní Initiative are very much needed in Ecuador. Ecuador depends on oil for a third of its national budget. The three oil fields in the Yasuní Initiative represent 20 percent of its crude oil reserves. Ecuador faces the prospect of exhausting its oil reserves in the 2020s.[22] Its economy is dependent on exports of oil and agricultural products and foreign remittances[23] from Ecuadorians living abroad. It has taken years to consolidate fiscal discipline.[24]

The estimated windfall of $350 million each year from the Yasuní Initiative could assist in raising the social indicators for Ecuador. According to the World Bank's 2009 database of Health, Nutrition and Population (HNP), poverty rates increased from 40 to 45 percent between 1990 and 2001.[25] In Latin America, only Bolivia and Guatemala are poorer. Ecuador ranks among countries with a medium level of human development, ranking 89 among the 177 countries measured, but national percentages of poverty and extreme poverty measured by consumption remain at levels close to those of 10 years earlier.[26] In 2005, Ecuador ranked 20 out of 21 countries in the region in social expenditure according to the Economic Commission for Latin America and the Caribbean (ECLAC), with a per capita social expenditure of $76.00 (USD) compared with the regional average of $610.00 (USD).[27] Although Ecuador has increased these expenditures over the last few years, it is still relatively low compared to the percentages invested by other Latin American countries into social programs like education, vocational training, and health care. The Yasuní Initiative could provide the funds necessary to reduce extreme poverty.

In addition to the poverty seen in urban areas, Ecuador suffers from chronic, structural poverty that is concentrated in rural areas, particularly indigenous and Afro-Ecuadorian communities.[28] Rural poverty was measured at 59.7 percent compared to 22.6 percent in urban areas.[29] Particularly in the Amazon region, the communities have a severe lack of access to basics like uncontaminated water and land, social programs and education available in urban areas. They are therefore largely excluded from agricultural and non-agricultural opportunities.[30] Poverty among indigenous people in Ecuador has reached 80.2 percent compared with 57.9 percent among non-indigenous groups.[31]

The Huaorani people occupy Amazonian regions of Ecuador and live primarily from hunting, fishing, gathering, and family farming. Main agricultural crops include yucca and banana, but in some areas, they can also grow sweet potatoes, guava, papaya, pineapples, and peanuts.[32] As well as being home to the Huaorani people, the Yasuní is also occupied by the Taromenane and Tagaeri peoples. After migrating to these remote regions to escape colonizers and other tensions, their lands were soon threatened by illegal logging and oil extraction prospectors.[33] In Coca, the capital of the Amazonian province of Orellana, medical practitioners report that after oil drilling began in the area, the Huaorani people began to suffer from gastrointestinal, respiratory and skin disorders, allegedly caused by the contamination of the River Tiputini.[34]

After years of exploiting these lands with valuable oil drilling, the economic indicators showed that living conditions and the economy did not improve. The Amazonian provinces of Ecuador produce oil, but are also the areas with the greatest poverty levels, with land which has been contaminated by waste from oil production. Legislation to protect the Yasuní National Park from drilling has been a strong goal supported by other countries in order to combat climate change, to reduce the destruction of biodiversity, to protect the Huaorani, Tagaeri and Taromenane peoples and to diversify Ecuador's economy.[35] Nonetheless, these groups are in danger of losing their livelihoods, and possibly their lives, if sustainable alternatives are not fostered.

The Gross National Income (GNI) per capita in Ecuador for 2008 was about $3,690.00 compared with $7,300.00 in Brazil, $9,200.00 in Venezuela, and substantially lower than the GNI for Columbia and Peru.[36] The manufacturing and agricultural sectors account for the majority of employment income for Ecuadorians, including banana and tuna exports and manufacturing automobile parts. According to numbers from the Banco Central del Ecuador, in the last two years, real oil GDP (Gross Value Added in oil production and the manufacturing of oil products) contracted at an annual rate of 7.6 percent, compared to a 4.7 percent growth rate between 2000 and 2006. In contrast, real non-oil GDP (Gross Value Added in production activities other than oil extraction and manufacturing of oil products) grew by 5.9 percent per year two years, compared to an annual rate of 4.8 percent for the 2000–2006 period.[37] It is estimated that 1,225,000 hectares were deforested as the result of 994 oil platforms and related road network, and that 7,148 barrels of spilled crude oil were not recuperated between 1994 and 2000.[38] Many parts of Ecuador's Amazon region have been covered by exploration or exploitation contracts, furthering the country's dependence on such traditional exports and further exposing its population to water and soil contamination.

The Yasuní Initiative could provide funds to grow sectors other than oil. Employment in the oil and mining sector represents only around 0.5 percent of total national employment.[39] Diversifying in the non-oil sector can arguably have a higher and more direct impact on employment, labor income and poverty reduction in an oil-producing country like Ecuador, where the construction, manufacturing, trade and repair sectors comprise around 37.9 percent of total employment, agriculture takes up 30.2 percent of employment.[40]

The Challenges concerning the Yasuní Initiative

The Yasuní Initiative faces several obstacles, primary among these is money. The Yasuní Initiative is to raise $350 million for each of ten years. Contributors to the Yasuní Initiative are to include governments and philanthropists. The Clinton Global Initiative named the Yasuní Initiative one of its Member Commitments. The Wallace Global Fund has also endorsed the plan,[41] and brought in Acción Ecológica, Amazon Watch, Earth Economics, University of Maryland, C.S. Mott Foundation, and the Pachamama Alliance. To date, no money has come in to the Fund.

Tentative commitments from the governments of Spain, Germany, Belgium, France, and Sweden to contribute $1.7 billion — close to half the amount demanded by President Correa — have been secured. Norway has also shown great interest in the plan. The German government initially agreed to pay $50 million (36 million Euros) into a future international fund,[42] later withdrawn by the German Ministry of International Development due to concerns about the plan itself.[43] It remains unclear what the German government will do.

It is not just the Western European industrialized countries that have been tapped by the Ecuadorian government for donations to the Yasuní Initiative. President Correa appointed former Ecuadorian foreign minister Maria Fernanda Espinosa to head a panel to seek donations from Arab and Asian countries for the Yasuní Initiative. The Vice-President of Ecuador has travelled to Iran, Turkey and Dubai to raise funds. Latin American countries are slowly coming in with support. On May 20, 2011, the government of Perú announced that it would contribute $300,000 to the Yasuní Initiative.[44] A contribution of $100,000 has been pledged by Chile.[45]

But conditions are being attached to the contributions. Prospective donors have demanded some control over how the money is spent and asked Ecuador to expand the amount of land protected from development under the initiative. President Correa has made a number of poor public relations moves, blasting donors, calling the conditions of potential donors "unacceptable" and "embarrassing" affronts to Ecuador's sovereignty. In addition, the members of the environmentalist advisory panel resigned in January 2010 after Ecuadorian President Rafael Correa publicly berated them, labeling them "infantile environmentalists."[46] Poor donor relations have had repercussions. With the signing of a Memorandum of Agreement between Ecuador and the United Nations Development Programme, a fund was set up by the UNDP's Multi-Donor Trust Fund Office to administer contributions. To date, there has been no money or probability of money for the Yasuní Initiative to be even partially funded. The Western industrialized states are now more focused on the financial health of Europe[47] and the future of the Euro zone.[48] Sovereign wealth funds and global financiers no longer have the disposable funds necessary to fund the Yasuní Initiative[49]. And questions are being raised about the fund's management and the Ecuadorian government's ability to keep to its end of the agreement — not to exploit oil in the Yasuní National Park. President Correa's threat of exploitation of the Yasuní-

ITT oil fields if the Yasuní Initiative is not funded hangs in the background.

Beyond the financial issues are complexities as to whether Ecuador can make the Yasuní Initiative work within its new constitutional framework. Ecuador's ability to develop other parts of the Yasuní National Park that are not part of the ITT fields is shaky at best. Would the donors have the right to confiscate the oil if it does end up being extracted from the ITT fields? What kind of guarantees can ensure that the Ecuadorians follow their end of the deal? The proposal would block drilling in three oil fields in the Yasuní National Park, but it does not explicitly prohibit development in the rest of the park. The government of Ecuador could still allow extractive industries access to the park. The initial Yasuní Initiative model, as presented by Ecuadorian environment minister Marcela Aguiñaga, was only meant for regions where petroleum reserves are located beneath extremely sensitive biodiverse ecosystems. This raises a further political question: should the monies earned be shared by all regions or just among the people who live in the Park itself?

Ecuador is struggling with new constitutional measures, further referenda,[50] and general unrest among the police.[51] The constitution of 2008 is highly activist and provides a number of confusing rights to a variety of interests concerning the extraction of natural resources. Under Ecuador's new constitution, no oil drilling is allowed in national parks, unless it is a "national priority." If a project is deemed a national priority, as the Yasuní Initiative is, then a national consultation process must be carried out, the terms of which remain unclear. Correa himself has indicated that if the proposal fails, he would call a referendum, asking Ecuador's public whether drilling should occur.

The constitution does not spell out a system of community voting for environmental management projects like the Yasuní Initiative. Article 398 stipulates that "[e]very decision or state authorization that may affect the environment should be consulted with the community, which will be informed extensively and opportunely." But this does not yet mandate a public vote before moving on with environmental management projects. Article 398 provides this merely consultative process rather than a process requiring consent, making the results of a consultation process non-binding. Article 407 provides for a popular referendum concerning possible extractive activity of non-renewable resources in the areas protected. Yet it is unclear who is to attend a public meeting, who is to vote, and whether 50 percent of the votes would be enough to pass such a measure. Indeed, the community would have to be properly informed ahead of time that they were free to challenge executive decisions through the judicial system. If popular resistance is visible upon informing communities, then there would be proper review by the Constitutional Court and other elected officials to make sure the environmental project was constitutional. This seems to be the extent of vague community consent needed. But it is still open to interpretation and litigation through Ecuadorian courts.

Indeed, a number of competing interests will require judicial clarification as management projects continue. Article 57 provides for collective rights for towns, communities and indigenous people to participate "in the use, administration and conservation of the renewable natural resources that be found in their lands." Yet the State continues to play a role in management issues through Article 404, a provision which states that the management of Ecuador's natural patrimony "will be held to the principles and consecrated guarantees in the constitution and will be carried out according to the territorial code and an ecological zoning, according to the law." Further, Article 405 provides that the conservation of the biodiversity and the maintenance of the ecological functions will be for the state to decide, and that the state will assign the necessary economic resources for the financial sustainability of the system of protected areas. For the Yasuní Initiative to come to fruition, these internal issues in Ecuador must be settled, most likely through the courts, rather than mandated by presidential decree.

Conclusion: The Yasuní Initiative and Creative Problem Solving

The Yasuní Initiative builds on the "debt for nature swaps" that were all the rage in the 1980s. These were mechanisms to assist developing countries which were having trouble paying back their debts — let alone servicing them — by exchanging conservation easements on the country's natural resource development to slow down the plunder of forests and other natural habitat. "The Initiative is an innovative response from Latin America to the procrastination of the global negotiations [on climate change]."[52] But critics called this a violation of state sovereignty — with wealthier states dictating the development policy of less developed countries.

The Yasuní Initiative is an important proposal for future development models — one where value is created by not drilling or taking down forests, but instead from not doing so.[53] Combining climate protection and the fight against poverty will play a growing role in the future international discussions: "Magic fixes for global warming are regularly proposed and make headlines in the media. Only long-term strategies and global measures, such as the increase of fossil fuel efficiency and the gradual worldwide reduction in emissions will enable us to meet climate targets."[54]

A number of creative schemes, like the Yasuní Initiative, have been developed to make fungible the release of carbon dioxide in the atmosphere. Brazil has developed an Amazon Fund, which seeks to attract money to conserve the rain forest.[55] Norway has pledged up to $1 billion over the next several years to combat climate change by protecting forests.[56] The Japanese government agreed to lend Peru $120 million to protect 212,000 square miles of Amazonian rain forest over the next decade.[57] These kind of redistributive, forward-thinking and preventive programs will, without doubt, be part of the overall multilateral and interdisciplinary solutions to global climate change that could assist in the sustainable development of all Amazonia.

Greenpeace estimates that $30 billion are needed to secure the future of the rainforests worldwide.[58] The Amazon basin, which covers part of eight sovereign states, contains 45 percent of earth's remaining rain forest. In the last 25 years, about 17 percent of the rain forest has been cut down. The Amazon stores between 80 billion and 130 billion metric tons of carbon in its leaves and trunks. If burned, this would emit 50 times more carbon dioxide than the annual output of carbon dioxide from the United States. Saving part of Ecuador's rain forest is a start. It can be a cheaper and faster way to avoid greenhouse gas emissions than replacing coal-fired power plants with renewable energy or switching to electric cars. But there is a real issue over how much of an appetite the industrialized world has for such projects, particularly in the wake of the global financial crisis.

Acknowledgments

The author thanks Yesenia Acosta and Cedric Anderson, Juris Doctor Candidates at California Western School of Law, for their research assistance for this chapter and Jaime Vintimilla for his advice on Ecuadorian constitutional law.

Notes

1 United Nations Development Programme Newsroom, *UNDP, Ecuador sign deal to protect Amazon from oil drill*, August 3, 2010, available at http://content.undp.org/go/newsroom/2010/july/PNUDyEcuadorsuscribenacuerdoparalainiciativaYasuni.en (last visited on Aug. 21, 2010). This is more than an earlier report that estimated that the initiative would prevent 410 million tons of carbon dioxide from being emitted. *See* Joshua Partlow, *In Ecuador, an Unusual Carbon-Credit Plan to Leave Oil Untapped*, WASH. POST, May 26, 2009, available at http://www.washingtonpost.com/wp-dyn/content/article/2009/05/25/AR2009052502402.html. For the Wallace Global Fund, "the proposal will avoid the emission of 407 million metric tons of CO_2 that would result from burning the extracted fossil fuels." Wallace Global Fund, Leaving Ecuador's Oil in the Ground, 2007, available at http://www.clintonglobalinitiative.org/commitments/commitments_search.asp?Section=Commitments&PageTitle=Browse%20and%20Search%20Commitments (last visited on May 27, 2011). The Economist reports it at 436 million tons. *Reailty Check*, ECONOMIST, Sept. 22, 2010, available at http://www.economist.com/blogs/americasview/2010/09/ecuadors_environment (last visited on May 27, 2011).

2 James Cooper & Janeen Kerper, *Creative Problem Solving*, in THE ESSENTIALS OF HUMAN RIGHTS (edited by Rhona K.M. Smith & Christien van den Anker; Hodder Arnold U.K.) 72 (2005).

3 For Thomas D. Barton, Creative Problem Solving aims "to make the law a more sensitive and respectful shaper of the social, physical and relational environment . . . [and] to give lawyers the understanding, skills, and attitudes needed to apply tools of persuasion and reconciliation where [those tools] may be more appropriate." Thomas D. Barton, *Creative Problem Solving: Purpose, Meaning and Values*, 34 CAL. W. L. REV. 273, 283–284 (1998). Jessica Fink, *Unintended Consequences: How Antidiscrimination Litigation Increases Group Bias in Employer-Defendants*, 38 N.M. L. Rev. 333 (2008). *See also* THOMAS D. BARTON, PREVENTIVE LAW AND PROBLEM SOLVING: LAWYERING FOR THE FUTURE (2010).

4 James Cooper, *Drug Treatment Courts: A Problem-Solving Mechanism to Address the Demand Side of the War on Drugs*, in JUSTICIA PENAL Y ADICCIONES: TRIBUNALES DE TRATAMIENTO COMO ALTERNATIVA A LA SANCIÓN 13–22 (Lexis Nexis Chile, 2007).

5 James M. Cooper, *Se abre una nueva oportunidad: Importando sistemas horizontales de justicia durante una época de reforma judicial*, 2 Revista CREA 96 (2001).

6 *See generally*, Barbara J. Cox, *AALS as Creative Problem-Solver: Implementing Bylaw 6-4(A) to Prohibit Discrimination on the Basis of Sexual Orientation in Legal Education*, 56 J. LEGAL EDUC. 22 (2006). Kimberlee K. Kovach, *Lawyers Must Keep Pace with Practice: Plurality in Lawyering Demands Diverse and Innovative Ethical Standards*, 39 IDAHO L. REV. 399 (2003). Stefan H. Krieger, *Domain Knowledge and the Teaching of Creative Legal Problem Solving*, 11 CLINICAL L. REV. 149 (2004). Linda Morton, *A New Approach to Health Care ADR: Training Law Students to be Problem Solvers in the Health Care Context*, 21 GA. ST. U. L. REV. 965 (2005). Annelise Riles, *A New Agenda for the Cultural Study of Law: Taking on the Technicalities*, 53 BUFF. L. REV. 973 (2005). Spencer Rand, *Teaching Law Students to Practice Social Justice: An Interdisciplinary Search for Help Through Social Work's Empowerment Approach*, 13 CLINICAL L. REV. 459 (2006).

7 James M. Cooper, *Creative Problem Solving and the Castro Conundrum*, 28 CAL. W. INT'L L. J. 391 (1998).

8 Rhona K. M. Smith, *The International Impact of Creative Problem Solving: Resolving the Plight of Indigenous Peoples*, 34 CAL. W. L. REV. 411 (1998).

9 Dino Kritsiotis, *The Power of International Law as Language*, 34 CAL. W. L. REV. 397 (1998).

10 BRIGID STARKEY, MARK A. BOYER, & JONATHAN WILKENFELD, NEGOTIATING A

COMPLEX WORLD: AN INTRODUCTION TO INTERNATIONAL NEGOTIATION (1999).

11 Richard Falk, *Tom Farer's Liberal World Order: A Realist Utopia*, in 9 HUMAN RIGHTS & HUMAN WELFARE: A Symposium on Confronting Global Terrorism and American Neo-Conservatism: The Framework of a Liberal Grand Strategy (2008) 5 (2009).

12 "Given its potential impact on ecosystems and human activities, climate change is recognized as one of the most important challenges for this century" in INTERNATIONAL TRADE REGULATION AND THE MITIGATION OF CLIMATE CHANGE (edited by Thomas Cottier, Olga Nartova & Sadeq Z. Bigdeli) 10 (2009).

13 See William J. Aceves, *Using Transnational Litigation to Protect the Amazon: A Study on Corporate Litigation, infra*.

14 *See generally* Suzana Sawyer, *Suing ChevonTexaco*, in THE ECUADOR READER: HISTORY, CULTURE POLITICS (edited by Carlos de la Torre & Steve Striffler) 321 (2008).

15 "Ecuador is substantially dependent on its petroleum resources, which have accounted for more than half of the country's export earnings and approximately one-third of public sector revenues in recent years." Central Intelligence Agency, Ecuador, World Factbook, available at https://www.cia.gov/library/publications/the-world-factbook/geos/ec.html (last visited on Nov. 10, 2011).

16 *Ecuador's Yasuní Initiative*, ECONOMIST, July 2, 2009, available at http://www.economist.com/world/americas/displaystory.cfm?story_id=E1_TPJVGPDS (last visited on April 23, 2010).

17 Graciela Chichilnisky, *Foreward: Yasuní: The New Economics of Planet Earth*, in JOSEPH HENRY VOGEL, THE ECONOMICS OF THE YASUNÍ INITIATIVE, xvii–xviii (2009).

18 Eric Marx, *Germany's Withdrawal of Funding Threatens Plan to Save Ecuador Forest*, SCIENCE *INSIDER, June* 9, 2011, available at http://news.sciencemag.org/scienceinsider/2011/06/germanys-withdrawal-of-funding.html (last visited on Nov. 10, 2011).

19 UNESCO, Biosphere Reserve Information: Ecuador, available at http://www.unesco.org/mabdb/br/brdir/directory/biores.asp?code=ECU+02&mode=all (last visited on April 25, 2010).

20 *See* Jens-Christian Svenning, *Microhabitat specialization in a species-rich palm community in Amazonian Ecuador*, 87 J. ECOLOGY 55, at 55 (1999) (reporting on the patterns of distribution and abundance in a high diversity palm community in terra firme rain forest in relationship to topography, soil drainage and canopy height at spatial scales of tens to millions of meters). In the Yasuní National Park, there are three main types of vegetation: "'Terra firme', found on the high reliefs areas and not subject to flooding; 'Varzea', a forest type subject to periodic flooding, and 'Irapo' in the permanent or near permanent flooded forest." UNESCO, Biosphere Reserve Information: Ecuador, available at http://www.unesco.org/mabdb/br/brdir/directory/biores.asp?code=ECU+02&mode=all (last visited on April 25, 2010).

21 Chichilnisky, *supra* note 17, at xv.

22 Dominque Hechette, *Trade Policy and Competition*, in ECUADOR: AN ECONOMIC AND SOCIAL AGENDA IN THE NEW MILLENNIUM (edited by Vicente Fretes-Cibils, Marcelo M. Giugale, & José Roberto López-Calix) 163 (World Bank, 2003), at 16.

23 Brad D. Jokisch & David Kyle, *Ecuadorian International Migration*, in THE ECUADOR READER: HISTORY, CULTURE, POLITICS (edited by Carlos de la Torre & Steve Striffler) 350 (2008).

24 THE WORLD BANK, CREATING FISCAL SPACE OF POVERTY REDUCTION IN ECUADOR: A FISCAL MANAGEMENT AND PUBLIC EXPENDITURE REVIEW 11 (2005).

25 The World Bank, Health, Nutrition, and Population, available at http://web.worldbank.org/WBSITE/EXTERNAL/TOPICS/EXTHEALTHNUTRITIONANDPOPULATION/EXTDATASTATISTICSHNP/EXTHNPSTATS/0,,menuPK:3237172~pagePK:64168427~piPK:64168435~theSitePK:3237118,00.html (last visited May 3, 2010).

26 United Nations Development Programme, *Assessment of Development Results in Ecuador*, 16 (2008) available at http://www.undp.org/evaluation/documents/ADR/ADR_ Reports/ ADR_Ecuador.pdf (last visited Nov. 6, 2011).

27 *Ibid.*

28 IFAD — Rural Poverty Portal, *La Pobreza Rural en Ecuador*, (2009) available at http://www.ruralpovertyportal.org/web/guest/country/home/tags/ecuador# (last visited May 4, 2010).

29 Mark Weisbrot & Luis Sandoval, *Update on the Ecuadorian Economy*, CERP, June 2009, available at http://www.cepr.net/index.php/publications/reports/update-on-the-ecuado-rian-economy/ (last visited May 4, 2010).

30 IFAD — Rural Poverty Portal, *La Pobreza Rural en Ecuador*, (2009) available at http://www.ruralpovertyportal.org/web/guest/country/home/tags/ecuador# (last visited May 4, 2010).

31 Building Equality and Opportunity through Social Guarantees: New Approaches to Public Policy and the Realization of Rights 143 (Estanislao Gacitua-Mario et al., eds.,World Bank 2009).

32 *Ibid.*

33 Matt Finer et al. Varsha Vijay, Fernando Ponce, Clinton N. Jenkins, and Ted. R. Kahn, *Ecuador's Yasuní Biosphere Reserve: a brief modern history and conservation challenges*, ENVIRON. RES. LETT. 4, 10 (2009).

34 Kintto Lucas, *From Pollution to Protection*, Global Knowledge No. 2 (2007) available at http://siu.no/magazine/Conferences-and-publications/Global-Knowledge/Issues/No-2-2007/From-Pollution-to-Protection (last visited May 4, 2010).

35 *Ibid.*

36 The World Bank, Health, Nutrition, and Population, available at http://ddpext.world-bank.org/ext/ddpreports/ViewSharedReport?&CF=&REPORT_ID=13609&REQUEST _TYPE=VIEWADVANCED&HF=N&DIMENSIONS=17 (last visited May 4, 2010).

37 Mark Weisbrot & Luis Sandoval, *Update on the Ecuadorian Economy*, CERP, June 2009, available at http://www.cepr.net/index.php/publications/reports/update-on-the-ecuado-rian-economy/ (last visited May 4, 2010).

38 United Nations Development Programme, *Assessment of Development Results in Ecuador*, 16 (2008) available at http://www.undp.org/evaluation/documents/ADR/ADR_Reports/ ADR_Ecuador.pdf (last visited Nov. 6, 2010).

39 Mark Weisbrot & Luis Sandoval, *Update on the Ecuadorian Economy*, CERP, June 2009, available at http://www.cepr.net/index.php/publications/reports/update-on-the-ecuadorian -economy/ (last visited Nov. 6, 2010).

40 *Ibid.* at 16.

41 Wallace Global Fund, Earth Economics Grants 2007, available at http://www .wgf.org/grants/2007/earth-economics (last visited on May 27, 2011).

42 Dow Jones, *Ecuador President: Germany Pledges $50 Million/Year For Yasuni Project*, available at http://www.nasdaq.com/aspx/stock-market-news-story.aspx?storyid=200910 270815dowjonesdjonline000239&title=ecuador-presidentgermany-pledges-50-million/ year-for-yasuni-project (last visited on Mar. 8, 2011).

43 Chris Lang, *Germany to pull out of Ecuador's Yasuni initiative?*, red-monitor.org (Sept. 22, 2010), available at http://www.redd-monitor.org/2010/09/22/germany-to-pull-out-of-ecuadors-yasuni-initiative/ (last visited on Mar. 8, 2011). *Cf.* Presidencia de la República del Ecuador, *Ministro Patiño aclara que no hay información oficial sobre el no aporte de Alemania al Yasuní*, Sept. 16, 2011, available at http://www.presidencia. gov.ec/index.php?option=com_content&view=article&id=185:ministro-patino-aclara-que-no-hay-informacion-oficial-sobre-el-no-aporte-de-alemania-al-yasuni-&catid=39:no ticias-importantes&Itemid=98 (last visited on May 27, 2011).

44 Presidencia de la República del Ecuador, *Perú contribuirá con $300 mil a la Iniciativa Yasuní ITT*, May 21, 2011, available at http://www.hoy.com.ec/noticias-ecuador/peru-aportara-300-mil-a-iniciativa-yasuni-itt-476453.html (last visited on May 12, 2012).

45 *Reality Check*, ECONOMIST, Sept. 22, 2010, available at http://www.economist.com/blogs/americasview/2010/09/ecuadors_environment (last visited on Oct. 27, 2011).

46 Associated Press, *Ecuador Minister Resigns Over Amazon Oil Project*, Jan. 15, 2010,

47 David Enrich & Laura Stevens, *Old Debts Dog Europe's Banks*, WALL ST. J., Nov. 7, 2011, at A1.

48 Charles Forelle, David Gauthier-Villars, & Marcus Walker, *Europe's Greece Ultimatum*, WALL ST. J., Nov. 3, 2011, at A1.

49 At the time of writing, there have been renewed commitments to the Yasuní Initiative — some $52 million in public and private funding promises. PR Newswire, *New Commitments Announced for Yasuni-ITT Campaign to Protect Ecuadorian Amazon at UN Meeting Hosted by UN Secretary General and President of Ecuador*, U.S. Politics Today, Sept. 27, 2011, available at http://uspolitics.einnews.com/pr_news/58977378/new-commitments-announced-for-yasuni-itt-campaign-to-protect-ecuadorian-amazon-at-un-meeting-hosted-by-un-secretary-general-and-president-of-ecuador (last visited on Nov. 11, 2011). This amount is still far less than the $300 million needed annually to fund the proposed project.

50 Jim Wyss, *Ecuador referendum tightens*, MIAMI HERALD ON LINE, May 10, 2011, available at http://www.miamiherald.com/2011/05/10/2210885/ecuador-referendum-tightens.html (last visited on May 27, 2011).

51 *Ecuador declares emergency as police protest, president is attacked*, CNN World News, Sept. 30, 2010, available at http://articles.cnn.com/2010-09-30/world/ecuador.violence_1_police-protest-national-police-police-headquarters?_s=PM:WORLD (last visited on Oct. 27, 2011).

52 Chichilnisky, *supra* note 17, at xv.

53 *See* JOSEPH HENRY VOGEL, THE ECONOMICS OF THE YASUNÍ INITIATIVE, xvii–xviii (2009).

54 Thomas Stocker, *Earth in the greenhouse — a challenge for the twenty-first century*, in INTERNATIONAL TRADE REGULATION AND THE MITIGATION OF CLIMATE CHANGE (edited by Thomas Cottier, Olga Nartova & Sadeq Z. Bigdeli) 10 (2009).

55 *Brazil launches rainforest fund*, BBC News, Aug. 1, 2008, available at http://news.bbc.co.uk/2/hi/7538480.stm (last visited on Nov. 7, 2011). *See generally* The Amazon Fund, available at http://www.amazonfund.gov.br/FundoAmazonia/fam/site_en (last visited on Nov. 7, 2011).

56 *Norway donates $1 billion to Brazilian rainforest fund*, Norway: The Official Site in the United States, Royal Norwegian Embassy in Washington, available at http://www.norway.org/ARCHIVE/policy/environment/regnskogen_i_brasil_en/ (last visited on Nov. 7, 2011).

57 *Japan loans Peru $120 million to conserve Amazon Rainforest*, ANDINA: AGENCIA PERUANA DE NOTICIAS, May 12, 2009, available at http://www.andina.com.pe/ingles/Noticia.aspx?id=z73GoRkNNII= (last visited on Nov. 7, 2011).

58 Logging in the Amazon, Greenpeace International, available at http://www.greenpeace.org/international/campaigns/forests/amazon/logging-in-the-amazon (last visited on April 23, 2010).

PART II

National Politics and Policies

5 Venezuelan Amazonia: At an Environmental Crossroad

Edgard Yerena and Anna Sartorio

Human and Natural Determinants

The Venezuelan Amazonia has important geographical and biological differences when compared to the non-Venezuelan Amazonia. Its main physical feature is its massive cover by the Guiana Shield geological province, characterized mostly by rocky hills and tabletop mountains, known as tepuis.[1] Gold, diamonds, and strategic minerals, are extracted from these formations. The humid climate encompasses a high hydropower potential.[2] The soils are poor of nutrients and very acidic (mostly between pH 3 and 4).[3] If the forest cover and the vegetal soil are removed, regeneration of the forest is highly improbable.

From a political, administrative and territorial point of view, the Venezuelan Amazonia is divided into two states: Bolívar and Amazonas; the state of Bolívar is generally known as Guayana, and the second is plainly known as Amazonas. For the purposes of this paper, we will refer to these two states as Amazonia (Figure 5.1, overleaf).

This part of Venezuela is home to the largest diversity of Venezuelan indigenous peoples; some of them having retained significant parts of their culture.[4] The human footprint on large portions of this region is not clearly evident, except for the northern strip next to the Orinoco River, at the edges of the Ciudad Guayana industrial complex.[5] Moreover, a large portion of the state of Amazonas has not been heavily impacted by the mainstream Venezuelan culture. The south of the Orinoco River is the Venezuelan equivalent of Siberia for Russia or of Alaska for the United States, a kind of hinterland or "backyard".

A recent estimate of the remaining forest shows that, the largest extensions and the largest biodiversity of Venezuela are concentrated in the south of the Orinoco River.[6] However, most of this land is "public land"[7] or in Venezuelan legal terms "*baldíos*". Property of these lands belong to the state, including tracks of land occupied by indigenous peoples.[8] Before reaching statehood, the state of Amazonas was "federal territory", which implied a strict dependency on the central government and military decisions. In 1951, the acting military government decreed that all visitors to indigenous areas had to obtain special permits. Its main objective was a slow and gradual

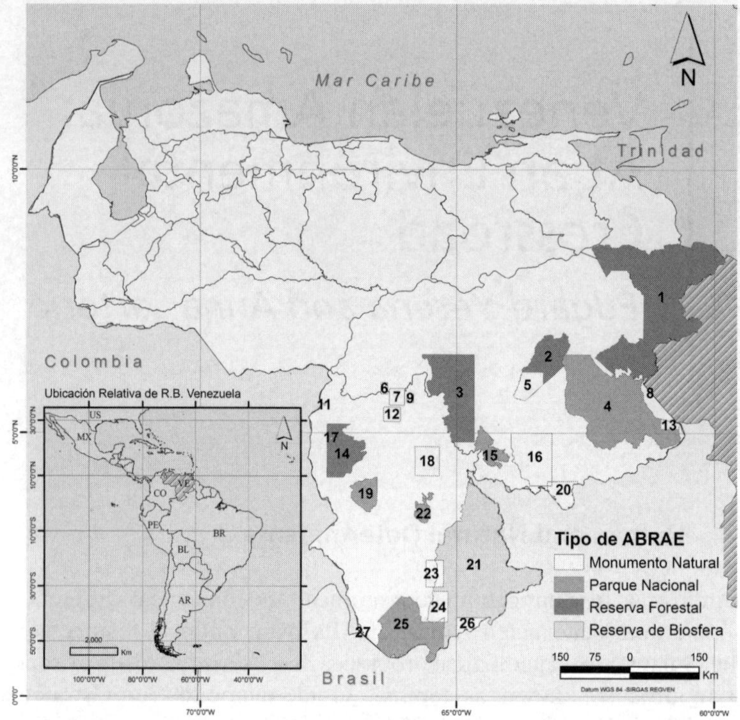

Referencia de ABRAE

ID	Nombre de ABRAE
1	IMATACA
2	LA PARAGUA
3	EL CAURA
4	CANAIMA
5	CERRO GUAIQUINIMA
6	CERRO GUANAY
7	SERRANIA YUTAJE-CORO CORO
8	CERRO VENAMO
9	CERRO YAVI
10	PIEDRA LA TORTUGA
11	PIEDRA PINTADA
12	CERRO CAMANI Y CERRO MORROCOY
13	CADENA DE TEPUYES ORIENTALES
14	SIPAPO
15	JAUA-SARISARIÑAMA
16	CERRO ICHUM Y CERRO GUANACOCO
17	CERRO AUTANA
18	MACIZO PARU-EUAJA
19	YAPACANA
20	SIERRA MARUTANI
21	ALTO ORINOCO-CASIQUIARE
22	DUIDA MARAHUACA
23	CERRO VINILLA Y CERRO ARATITIYOPE
24	SIERRA UNTARAN
25	SERRANIA LA NEBLINA
26	CERRO TAMACUARI-SERRANIA TAPIRAPECO
27	PIEDRA DE COCUY

Figure 5.1 Map of the Venezuelan Amazonia showing some of the areas under special management (listed in Table 5.1). Drawn from data provided by Ministerio del Ambiente.

incorporation of indigenous peoples to the national mainstream, while at the same time preventing the colonization by the "criollos" or the creole or mestizo population.[9]

Water: A Strategic Resource

Venezuela's water and energy come from Amazonia. Two of its watersheds (the Caroní and Paragua Rivers) in the State of Bolívar currently provide more than 73 percent of the electricity consumed in the country,[10] mainly produced by the Guri Dam. The region contains several other rivers with a high hydropower potential. In the early 1960s, the recently established democracy set out to push the growth of Venezuela's economy augmenting the supply of electricity, planning and investing in the building of large dams along the Caroni river watershed. With more hydroelectric generated power, Venezuela was able to export a significant amount of oil barrels, otherwise necessary as conventional fuel for power plants. It has been estimated that 282,000 oil barrels a day are needed to produce the equivalent amount of energy that the Guri Dam produces.[11] In terms of Venezuela's balance of trade, the dams' policy represented an income gain.

Table 5.1 Areas Under Special Management in the Venezuelan Amazonia

Category (translation from Spanish) and Name	State	Size in ha	Year of Designation
NATIONAL PARK (*Parque Nacional*)			
Duida-Marahuaca	Amazonas	210,000	1979
Parima-Tapirapecó	Amazonas	3,500,000	1991
Serranía La Neblina	Amazonas	1,360,000	1979
Yapacana	Amazonas	320,000	1979
Canaima	Bolívar	3,000,000	1962
Jaua-Sarisariñama	Bolívar	330,000	1979
NATURAL MONUMENT (*Monumento Natural*)			
Cerro Autana	Amazonas	30	1979
Piedra de Cocuy	Amazonas	15	1979
Piedra La Tortuga and Piedra Pintada	Amazonas	2,000	1992
Tepuyes -25 tracts-	Bolívar and Amazonas	1,750,000	1991
HYDRAULIC RESERVE (*Reserva Hidráulica*)			
Cuenca Alta del Ikabarú	Bolívar	190,000	1992
BIOSPHERE RESERVE (*Reserva de Biófera*)			
Alto Orinoco-Brazo del Cariquiare	Amazonas	8,400,000	1991
PROTECTION ZONE (*Zona Protectora*)			
Río Cataniapo	Amazonas	153,800	1993
La Tortuga Arrau	Bolívar	9,856	1989
Sur del Estado Bolívar	Bolívar	7,262,358	1974
FOREST RESERVE (*Reserva Forestal*)			
El Caura	Bolívar	5,134,000	1968
Imataca	Bolívar	3,203,250	1961
La Paragua	Bolívar	782,000	1968
Sipapo	Amazonas	1,215,500	1963
FOREST LOT (*Lote Boscoso*)			
Altiplanicie de Nuria	Bolívar	171,790	1965
San Pedro	Bolívar	757,400	1975
Caño Blanco	Bolívar	20,000	1984
CVG El Frío	Bolívar	65,000	1985
Parguaza	Bolívar	65,700	1986
El Dorado-Tumeremo	Bolívar	78,993	1987
Fundo Paisolandia	Bolívar	8,101	1988
Fundo Flamerich	Bolívar	19,196	1988

Sources: (a) Ministerio del Ambiente y de los Recursos Naturales Renovables — Areas Naturales Protegidas de Venezuela, 1992; (b) Ministerio del Ambiente y de los Recursos Naturales, Primer Informe de Venezuela sobre Diversidad Biológica. Caracas, June 2000.

Land Use Planning

Another strategy used to prevent Creole colonization of Amazonia, was to set aside large areas for special management or administration, known by the acronym ABRAE[12] (Table 5.1).

Two additional protection decrees of major importance were: N° 269[13] which prohibits mining activities in the Amazonas State and N° 1742[14] that also prohibits mining in the ABRAE Zona Protectora Sur del Estado Bolívar. The "Forest Reserve" figure is a legal protection status with two alternative objectives. The first one favors commercial and rational extraction of timber and other forest products.[15] The second one can be derived from historical facts, namely to postpone forestry and other activities awaiting a future decision on land use policy.[16] In all of the designated Forest Reserves, except Imataca (east of the State of Bolívar) the second objective applied. The "Forest Lot" figure is mainly focused on the first objective. "Hydraulic Reserve" was originally intended to discourage any land use that could jeopardize the Caroni River hydroelectric system, but in fact it has been loosely enforced. "Biosphere Reserve" follows international prescriptions of UNESCO, protecting traditional cultures and biodiversity at the same time. "Protection Zone" is a multiple land use scheme where special regulations should be applied to fulfill some specific environmental purposes. "National Parks" and "Natural Monuments", although explicitly intended to protect biodiversity, landscapes and natural features, also serve as safe places for indigenous peoples' territories and habitat.

The following administration of president Rafael Caldera implemented a controversial project called CODESUR, also known as "Conquest of the South" (1969).[17] Caldera's administration aligned itself with Brazilian development policies seeking to integrate both countries' developmental objectives, starting by surveying the natural resources potential of the entire region.[18]

The Gran Sabana subregion, largely within the Canaima National Park, was opened in 1972 by a road that interconnected industrial Ciudad Guayana with the Brazilian city of Boa Vista, across the border, and the National Park. With president Carlos Andrés Pérez (1974–1979) came the first national "environmentalist" policy.[19] He created the Environment Ministry, dismantled the CODESUR project, and prohibited commercial logging in the Amazonas territory.[20]

In yet another unprecedented decision, president Perez's administration designated two national monuments (the first ones in the Amazonia, year 1979, see Table 5.2) in order to protect outstanding tepuis and granitic rock features, as well as four national parks (year 1979, see Table 5.2), totaling 2,220,000 has, one of them (Serranía de la Neblina) contiguous to a smaller one previously established by the Brazilian government on its side of the border.

Mining

In the late 1970s and throughout the 1980s, the indigenous rights movement took a decisive momentum. This movement was mainly promoted through the action and representation of religious groups, anthropologists, and environment conservationists[21], and not so much by indigenous peoples themselves.[22] In the 1980s, the nomadic gold miners, mainly in east Guayana, became organized with sophisticated logistics and equipment. With political and financial support from their headquarters in the city of Boa Vista, these miners, known as "garimpeiros", started penetrating Venezuela's Amazonas state. [23] The result was an increased conflict with indigenous communities, especially the Yanomami. The Venezuelan government responded by completely banning the miners' activities and expelling them from the country, albeit not always with success. Such conflict resulted in the killing of sixteen[24] Yanomami Indians, in the so-called Haximú (Hashimo-teri) Massacre of 1993.[25] After that conflict, Venezuelan public opinion favored the strict conservation of Amazonas state.

During his second administration (1989–1993), president Pérez proposed a second set of protected areas as a political response to the garimpeiros' intrusion. In 1991 he established large tracts of land under the status of "Natural Monument" in order to preserve all of the tepui mountains, the Parima Tapirapecó "National Park", and a 84,000 km2 "Biosphere Reserve", the latter two on Yanomami territories at the Orinoco River headwaters. These protected areas were intended to become a safe environment and habitat for indigenous peoples and their sacred places, while at the same time, adding to biodiversity preservation. Despite such efforts, however, environmental law enforcement in these remote areas proved to be difficult.[26]

Conservation in Retreat

By 1992, democracy was seriously challenged by a failed military coup led by Hugo Chávez. Voters in Venezuela reacted by electing former president Rafael Caldera for a second mandate. His administration launched a sort of new CODESUR project in 1994, this time attached with a "sustainability" discourse[27] aligned with Brazil's geopolitical interests. [28] To further increase the provision of electricity to the Brazilian city of Boa Vista, an ambitious project was implemented to build a transmission line linked to the Caroni hydropower system. The project sparked strong reactions among indigenous communities as well as among environmental organizations, civil society, public and private institutions. [29] The main purpose of the transmission line was to provide energy to large-scale mining projects along its path to the city of Boa Vista in Brazil, barely considering the urgent needs of additional electricity for Venezuela itself. The project overlooked legal requirements and was considered illegal and unconstitutional, while also generating a considerable environmental and social impact.[30] The most significant issue in this period was the attempt to legalize industrial mining within the Imataca Forest Reserve, through a special land use regulation[31] which turned into one of the major environmental and political battles in the history of Venezuela.[32] This attempt was stopped when the Supreme Court ruled a precautionary decision ordering the non-enforcement of this regulation passed by a lawsuit presented by environmental organizations and concerned citizens.[33]

The general political crisis weakened the major political parties and the leader of the failed military coup attempted in 1992 was elected president of the Republic in 1998, promising to give priority to the protection of water resources and forests above mining ventures. [34] Shortly thereafter, a new Constitution (1999) with explicit provisions regarding environmental and human rights, citizen participation and indigenous peoples' rights, was enacted. Notwithstanding these provisions, however, the new administration decisively supported the expansion of the power line into Brazil, which went through the land of indigenous peoples within the Canaima

National Park, listed under the World Heritage List by Unesco in 1994,[35] despite the opposition. This project ignited the first environmental political conflict with the new president. [36] Also for the first time in modern times, clashes occurred between indigenous communities opposing the construction and the military forces, resulting in the killing of one member of the Pemon tribe.[37] After the confrontations, some Pemon leaders agreed to negotiate with the president with the promise that the government would formally recognize their ancestral land and their habitat territories. [38] Such a promise, however, was in stern contradiction to Venezuela's military doctrine of not recognizing any "territory" belonging to indigenous peoples, because of some fear of 'political separatism'.[39]

On the other hand, the current Venezuelan government, composed of many active and retired military officers, is said to have been secretly negotiating with Guyana`s government (former British Guyana) over the recognition of Guyanese sovereignty over territories that Venezuela has claimed for more than 100 years in the Essequibo river basin. The Guyanese government needs this recognition in order to be able to develop large mining projects as well as a hydroelectric dam.

In tandem, some important social changes are underway in the state of Amazonas. New political actors are taking key positions, specifically Colombian insurgent groups who provide protection to illegal gold miners,[40] mostly Creole Brazilian, Guyanese, Dominicans and Colombian citizens- in exchange for their participation in the profits within Venezuelan territory. Over the last five years, the Venezuelan government has taken actions to expel informal miners from the hydropower watersheds, alleging that such activities jeopardize the life span of turbines and other facilities. A 2,100 meter military airstrip was built in the center of the state of Amazonas, at La Esmeralda, in order to secure the frontier.[41]

The government has also decided to end scientific and technical international cooperation alleging international intrusion to Venezuela's Amazonia sovereignty.

All such actions have occurred under a veil of almost secrecy, a pattern increasingly present in most activities with an official governmental presence in the region, amidst accusations of severe human rights violations.[42] Indigenous participation in small scale gold and diamond mining was known both in the states of Bolívar and Amazonas, but a significant increase in these activities is currently observed.[43] Despite the initial electoral promises, industrial and small-scale mining was again allowed and legalized within the Imataca Forest Reserve, with a new decree[44] that left the earlier Supreme Court ruling without effect.[45] This time a lawsuit against the new decree was politically nonviable.[46] Moreover, the government took actions to shift the control over industrial gold mining into the hands of new international actors, especially companies from Russia, Iran and China, countries which are political allies of the current regime.

A pharaonic trans-amazonian pipeline project designed to transport natural gas from Venezuela to Argentina, surrounded by uncertainty, was also proposed by Venezuelan authorities.[47] A sentiment against non-Catholic missionaries developed in certain political and academic groups since the mid 1980s, based strongly on anti-U.S. attitudes. These attitudes turned into political action when president Chávez expelled the New Tribes Mission in 2005, leaving some indigenous communities, such as those of the Hoti tribes and some Yanomami communities, without the buffer barrier exerted for decades by these missionaries.[48] Moreover, informal miners and loggers are pushing into indigenous habitat and territories where they had not been previously reported, especially in the Caura River basin.[49]

At a Crossroad

Some general patterns may be detected based on the above facts. In general, there has been a cyclical shift between conservationism and conventional developmentalism. However, a conservationist tendency prevailed toward the state of Amazonas and the western portion of the state of Bolívar in spite of the fact that for a long time there had been a lawless and permissive tendency East of Bolívar, regarding diamond and gold mining, probably because of its vast expanses of 'untouched' forests and occupation by indigenous people.

The current tendency is a deepening lawlessness and an increasing complexity of the political situation in East Bolivar, which together with the stress caused to the indigenous communities, are spreading throughout the entire region. At the same time, the Venezuelan government shows a growing opacity and secrecy on its activities and decision making. What is even more striking is that the formal delimitation and recognition of indigenous peoples' lands seem to be politically non viable because of military opposition. If the current political conditions persist, environmental and social anarchy, conflict and disruption may well extend to the entire Venezuelan Amazonia, induced by old and new political actors.

Acknowledgments

The authors would like to give special thanks to Christine Hunefeldt, James Cooper and Hector Maldonado for inviting Edgard Yerena to present this paper in the "Amazonia: National Politics and Interests — Environment and Law" conference, held in San Diego, California, May 6–7, 2010. The authors also thank our Venezuelan colleagues Jorge Padrón, José Ramón Delgado, Charles Brewer-Carías and Alexander Luzardo, who provided useful comments and data.

Notes

1 Ministerio del Ambiente y de los Recursos Naturales, *Primer Informe de Venezuela sobre Diversidad Biológica* (2000).
2 Marta Miranda et al., "All That Glitters Is Not Gold: Balancing Conservation and Development in Venezuela's Frontier Forests", World Resources Institute (1997), available at: http://www.wri.org/publication/all-that-glitters-is-not-gold (last visited on Feb. 21, 2012).
3 Nelda Dezzeo, "Ecología de la Altiplanicie de la Gran Sabana", *Scientia Guaianae*, 4 (1994): 1–205.
4 Erika Wagner, "Población Indígena" en Imagen de Venezuela: Una Visión Espacial (Ediciones Petróleos de Venezuela) (1992).
5 Observatorio Mundial de Bosques, "Situación de los Bosques en Venezuela", *Fundación Polar* (2002).
6 Ibid.
7 Oficina Central de Estadísticas e Informática (OCEI), "El Censo Indígena de 1992", Caracas, Venezuela (1993).
8 Cuba Hernández, "Reforma Agraria y Tenencia de la Tierra en las Comunidades Indígenas del Amazonas Venezolano", *Amazonas Modernidad en Tradición* (1995): 227–246.
9 By "Creole" we refer to non-indigenous people.
10 Electrificación del Caroní CA (EDELCA) Río Caroní. Venezuela (1993).
11 Senator Reinaldo Rincón, "Declaraciones", *El Universal*, 2–18, Mar. 13, 1995.
12 Areas Bajo Régimen de Administración Especial.

13 Venezuela. "Decreto N° 269", Gaceta Oficial de la República de Venezuela, No 4.106 Extraordinario, June 9, 1989.

14 Venezuela. "Decreto N° 1742", Gaceta Oficial de la República de Venezuela, No. 31.783, Aug. 8, 1991.

15 Ministerio del Ambiente y de los Recursos Naturales Renovables, "Plan del Sistema Nacional de Areas Protegidas: Marco Conceptual", *Serie de Informes Técnicos* (1985).

16 Such policy is implicit under the term "reserve" according to its meaning of "setting aside for future use."

17 Comisión Presidencial para Asuntos Fronterizos Colombo-Venezolanos, "Amazonas. Diagnóstico y Estrategia de Desarrollo Fronterizo", Gráficas Monfort (1992): 233.

18 Sergio Benacchio, "El Desarrollo Agrícola en la Región Amazónica Venezolana", *Amazonia: Investigación sobre Agricultura y Uso de Tierras. Memorias de Conferencia*, Centro Internacional de Agricultura Tropical (1982): 117–138, 448.

19 Ley Orgánica del Ambiente, Gaceta Oficial de la República de Venezuela, No. 31.004, June 16, 1976.

20 Venezuela. "Decreto No 2552", Gaceta Oficial de la República de Venezuela, No. 31.408, Jan. 19, 1978.

21 Alexander Luzardo, "Amazonas, El Negocio del Siglo", Ed. Centauro (1988).

22 Miguel Angel Perera, "El Desarrollo Sustentable ¿Base de nuevos males o posibilidad de crecimiento etnoeconómico?," *Amazonas, Modernidad en Tradición*, Antonio Carrillo & Miguel Angel Perera (eds.), (GTZ-CAIAH-SADA AMAZONAS) (1995): 9–42.

23 Francisco Simancas, "Geopolítica Brasilera, Amazonia y Garimpeiros", Ediciones del Congreso de la República (1993): 53.

24 James Brooke, *Brazilians Reduce Indian Death Toll*, N.Y. Times, A4, Sept. 1, 1993.

25 Patrick Tierney, "Darkness in El Dorado" (2000); La Era Agrícola, "A un año de Haximú", 19 (1994), available at: http://www.eraecologica.org/revista_19/era_agricola_19.htm?notas_fundacion.htm~mainFrame%20y%20La%20Era%20Agr%C3%ADcola%20N#%2019 (last visited on Feb. 21, 2012).

26 Marcus Colchester and Fiona Watson, "Venezuela: Violations of Indigenous Rights", Report to the International Labor Office on the Observation of ILO Convention 107, (1995): 42, available at: http://www.wrm.org.uy/publications/index.html#briefings (last visited on Feb. 21, 2012).

27 Proyecto de Desarrollo Sustentable del Sur (PRODESSUR), Imago Centro Gráfico S.A. (1994).

28 Memorando de Entendimiento para el Suministro de Energía Eléctrica Venezuela — Brasil, Jan. 29, 1997.

29 Comisión Permanente el Ambiente y Ordenación Territorial del Senado de la República, Tendido Eléctrico a Brasil, Informe, Aug. 1998.

30 Ibid.

31 Venezuela. "Decreto N° 1850", Gaceta Oficial de la República de Venezuela, No. 36.315, May 28, 1997.

32 Julio C. Centeno, "La Fiebre del Oro Amenaza a Imataca", *La Era Ecológica*, 3 (2007), available at: http://www.eraecologica.org/revista_03/lee_03.htm?imataca.htm~mainFrame (last visited on Feb. 21, 2012).

33 The Supreme Court ruled on Nov. 13, 1997 ordering an injunction on concessions, contracts or authorizations for mining purposes in the Forest Reserve of Imataca.

34 Hugo Chávez Frías, "Programa de Gobierno", presented at Hotel Caracas Hilton on May 22, 2000, available at: http://www.analitica.com/bitblioteca/hchavez/programa2000.asp (last visited on Feb. 21, 2012).

35 United Nations Educational, Scientific, and Cultural Organization (UNESCO), "Canaima National Park", World Heritage Convention, available at: http://whc.unesco.org/en/list/701 (last visited on Feb. 21, 2012).

36 El Universal, *Sobre Corrupción se Alza Tendido Eléctrico*, 1–10, Oct. 17, 1999.
37 Silviano Castro, personal communication, Pemon indigenous leader of the community of San Rafael de Kamoirán (2001).
38 Vicepresidencia de la República, "Puntos de Entendimiento entre el Ejecutivo Nacional y las Comunidades Indígenas del Estado Bolívar para la prosecución de la obra de Sistema de Transmisión de Energía Eléctrica al sureste de Venezuela", Comisión Para la Atención de los Asuntos Indígenas, July 21, 2000.
39 Former senator Alexander Luzardo, personal communication, April 30, 2010.
40 Jodie Twose, "Venezuela: An Amazonian Adventure", Far and Further: Site for the Independent Traveler, *available at*: http://www.farandfurther.com/venezuela-travel/amazon-adventure.htm (last visited on Feb. 21, 2012).
41 Angelica Mora, "Chávez, Iraníes y Uranio", La Nueva Cuba, Feb. 23, 2005, available at: http://www.lanuevacuba.com/archivo/angelica-mora-9.htm (last visited on Feb. 21, 2012).
42 Marcos D. Valverde, "Arrecian denuncias contra Plan Caura", *Correo del Caroní*, July 2, 2010, available at: http://www.correodelcaroni.com/index.php?option=com_wrapper&view=wrapper&Itemid=174&id_articulo=156418 (last visited on Feb. 21, 2012); Armando Arevalo Bravo, *Ratifican sentencia por Masacre de La Paragua*, Correo del Caroní, Aug. 21, 2009, *available at*: http://www.correodelcaroni.com/component/option,com_wrapper/Itemid,174/?id=134306 (last visited on Feb. 21, 2012).
43 Charles Brewer-Carías, personal communication, renowned explorer of the Venezuelan Amazonia, May 3, 2011.
44 Decreto No 3.110. Gaceta Oficial de la República de Venezuela No 38.028. September 22, 2004.
45 Julio César Centeno. "Decretado El Saqueo de Imataca". http://www.eraecologica.org/keops/saqueo_imataca.htm
46 Former senator Alexander Luzardo, personal communication. April 30, 2012.
47 Joe Wessley, *South American Leaders Propose Massive Transamazonian Gas Pipeline from Venezuela Through Brazil and Argentina*, The University of New Mexico Institutional Repository — LABD, Mar. 3, 2006, available at: https://repository.unm.edu/dspace/bitstream/1928/10747/1/SOUTH%20AMERICAN%20LEADERS%20PROPO%20MASSIVE%20TRANSAMAZONIAN%20GAS%20PIPELINE%20FROM%20VENEZUELA%20THROUGH%20BRAZIL%20AND%20ARGENTINA.pdf (last visited on Feb. 21, 2012).
48 Diario Berea, "Venezuela: Los Yanomami Denuncian Abandono Tras Expulsíon de Misioneros Evangélicos", Oct. 15, 2008, *available at* http://ceirberea.blogdiario.com/1224085860/ (last visited on Feb. 21, 2012).
49 Carolina Maffia, "Vuelven a Encenderse Alarmas por Mineros en el Caura", *Correo del Caroní*, Aug. 3, 2010, available at: http://www.correodelcaroni.com/index.php?option=com_wrapper&view=wrapper&Itemid=174&id_articulo=158776 (last visited on Feb. 21, 2012).

6 Protecting Mother Earth in Bolivia: Discourse and Deeds in the Morales Administration.

Nancy Postero

Introduction: The World People's Summit

In April 2010, Bolivian President Evo Morales held an international conference on Global Climate Change near Cochabamba, Bolivia. Representatives, indigenous groups, and social activists from 140 countries attended "The World People's Summit on Climate Change and the Rights of Mother Earth" (also known as "the Cochabamba Summit"). Bolivia is already experiencing devastating effects of climate change, as Morales and his climate negotiating team have described on many occasions. Bolivia's Amazonian regions have suffered terrible flooding over the last few years, while its desert lowlands have witnessed severe droughts. In the highlands, the two main glaciers that provide drinking water are shrinking. The Chacaltaya glacier disappeared completely this year; others have already lost 40–50 percent of their capacity (Democracy Now 4/21/10; Rosenthal NYT 2009). So it was with a great sense of urgency that Morales convened the Cochabamba Summit.

In a festive air of popular democratic participation, 30,000 people and seventeen working groups met to negotiate resolutions intended to address the problems left unanswered by the "official" climate conference at Copenhagen some six months before. In his opening address, President Morales set the tone for the meeting:

> We are here because in Copenhagen, the so-called developed countries failed in their obligation to provide substantial commitments to reduce greenhouse gases. We have two paths: either Pachamama or death. Either capitalism lives or Mother Earth lives. Of course, brothers and sisters, we are here for life, for humanity, and for the rights of Mother Earth. Long live the rights of Mother Earth! Death to capitalism! (Democracy Now 4/21/10)

Morales had spent the previous year trying — unsuccessfully — to make these points. Morales began his crusade at the United Nations (U.N.) in September 2009, saying that he believed defending Mother Earth had become more important than defending human rights. He said that "Mother Earth, Planet Earth, can exist without human life, but human life cannot exist without Mother Earth" (Morales 2009). Then, at the Copenhagen Summit (United Nations Climate Change Conference) in December 2009, Morales stunned the summit by calling on world leaders to hold temperature increases over the next century to just one degree Celsius (instead of the two degrees Celsius finally agreed upon). Even more provocatively, he argued that rich countries should pay climate reparations — what he terms a "climate debt" — to those poorer countries suffering the effects of climate change. Warning of a "climate holocaust" that will destroy Africa and many island nations, he called for an international climate court of justice to prosecute countries for climate "crimes" (Vidal, 2009).

Although it did not receive much press in the mainstream media in the United States (U.S.), the conference was extremely productive. Progressive observers such as Democracy Now, Naomi Klein, and World Social Forum organizer, Boaventura de Sousa Santos, called attention to the democratic nature of the conference where ideas were debated by activists, scientists, and policy makers from around the world in contrast to the closed-doors approach of the U.S. and the U.N. (DN 4/21/10). Klein suggested that "Bolivia's enthusiastic commitment to participatory democracy may well prove the summit's most important contribution" (Klein, 2010). But the summit also produced some important new demands. Announced on the last day, the "Peoples Agreement" denounced the Copenhagen accords, and supported Morales' call to limit global warming to one degree Celsius. It also called for the passage of a Universal Declaration of the Rights of Mother Earth, the creation of an International Tribunal to prosecute polluters, the protection for climate migrants, the establishment of an "Adaptation Fund" to help countries affected by climate change, and the full recognition of the U.N. Declaration on the Rights of Indigenous Peoples (General Assembly Resolution 61/295 adopted on September 13, 2007), whose lands and livelihoods are most affected by climate change. The agreement concludes with the need to hold a global referendum to consult with the world's peoples on all of these issues, combining the fate of the planet with the need for global democracy.

Saving Pandora?

The Cochabamba Summit cemented Morales' international reputation, giving him a highly visible platform for his longstanding critiques of imperialism, militarism, and neoliberalism. One of the most important accomplishments of the summit, he told Democracy Now's Amy Goodman, was that instead of just talking about the effects of climate change, this summit examined the underlying cause, namely capitalism (DN 4/27/10). Morales has often spoken about the need to end irrational consumer-driven industrialization, saying that such forms of capitalism are the "worst enemy of humanity" (DN 12/17/09). To this Western form of development, Morales poses a liberatory alternative: a sustainable model of development based on indigenous values and reverence for the *pachamama*, or Mother Earth. The new Bolivian constitution enshrines as a foundational value the indigenous notion of *vivir bien*, or to live well. Morales often refers to this model, arguing that the only way to end global warming is to end the "search for living better," or what he characterizes as the goal of consumerist capitalism:

> Living better is to exploit human beings. It's plundering natural resources. It's egoism and individualism. Therefore, in those promises of capitalism, there is no solidarity or

> complementarity. There's no reciprocity. So that's why we're trying to think about other ways of living lives and living well, not living better . . . Living better is always at someone else's expense. Living better is at the expense of destroying the environment. (DN 12/17/09)

This environmental discourse has been enormously influential, at least at the symbolic level. The Peoples Agreement appears to have adopted this model, implying that the application of the Bolivian model at the international level is the only way to save the earth from the disaster looming on the all-too-near horizon. In August 2009, the United Nationals general assembly declared Morales a "World Hero of Mother Earth" (Munkton, 2009). The Buenos Aires Herald posted a graphic the following January, after the release of James Cameron's 3D blockbuster film Avatar, depicting a blue-hued Morales (like the indigenous Na'vi) paired with the film's hero. Morales went to view the film in La Paz (only his third movie ever) and said he identified with the film's "profound show of resistance to capitalism and the struggle for the defense of nature" (AP 1/12/10).[1] While this discourse has increased Morales' reputation abroad, at home in Bolivia, there are serious controversies over development.

In this chapter, I describe several development conflicts and analyze them in light of what some scholars are calling the 'new extractivism in Latin America'. These controversies highlight the fundamental tensions within the Movimento Al Socialismo ("MAS") project. On one hand, Morales and his government have vowed to put into place a sustainable development model based on indigenous values. On the other, the government continues to exploit its natural resources to bring in the income necessary to redistribute the benefits to Bolivia's poor populations. How this tension is resolved will be critical for both Bolivia's environment and her peoples.

The View From Bolivia

Naturally, Morales' international public performances have focused the spotlight on Bolivia's ecological record, and brought out all kinds of protesters, as people used the occasion to critique the Morales government, pushing it to put its "save the planet" discourse into action at home. At the summit, opposition groups joined together at Mesa 18 (the 18th workshop) to condemn Bolivia's failures. Mesa 18 was organized by the National Council of Ayllus and Markas of Qullasuyu (CONAMAQ), a well-known organization that represents indigenous nations in the Bolivian *altiplano* (highlands) (Cooper, 2010). The goal of the Mesa 18 participants was to show that the indigenous-led Morales government was "adopting the same broken capitalist development strategies of the past and risking the same damaging environmental impacts" (Cooper, 2010).

Bolivia has suffered terrible damage over the years from a developmental model based on the extraction and exportation of natural resources. In 1996, a tailings dam broke near the mining center of Potosí, flooding the Pilcomayo River with toxic materials. Because the mine was owned by then President Sánchez de Lozada, the government did not insist on remediation (Farthing, 2009:27). Environmentalists and local activists are still pushing for remedies, and calling to close down the many ore-processing plants in the region (Ibid). In 2000, a pipeline broke near a shallow part of Lake Titicaca, spilling thousands of gallons of oil into the Desaguadero River, contaminating 2,400 square miles of crop and grazing lands belonging to local indigenous people (Farthing, 2009:26, Haglund, 2008). Again, little has been done to assist the victims. Indeed, Bolivia has a less than enviable record in terms of environmental protection.

Mines and Roads

The issue that received the most coverage at the Cochabamba summit was the San Cristóbal mine, in the Nor Lípez province of Potosí department, near the Salar de Uyuni. The Japanese corporation, Sumitomo, has operated the massive open pit mine since 2008, when it took it over from Apex Silver, exploiting silver, lead, and zinc. Over the weeks leading up to the summit, the Regional Federation of Peasant Workers of the Southern Altiplano (FRUCTAS) led community members and workers in protest, blockading the roads and taking over the offices (López Pardo, 2010: DN 4/22/10). FRUCTAS representatives say the mine uses 15,000 cubic meters of water a day — that's 600 liters every second — from the aquifer below the mine without paying for it (ibid.). The result, they say, is that local streams have dried up, the rivers are polluted with toxic refuse, and agriculture has been ruined (DN 4/22/10). They demand that the mine replenish the massive amounts of water it is currently using in its extraction processes and that the usage be subject to taxes. The mine claims it has lived up to its obligations under the 1997 mining law, and that it has paid about $350 million dollars in taxes over the last 10 years, although nothing for water (Carvajal, 2010). The community members also want the mine to fulfill the promises it made when it began operations, such as electrification and improved roads (López Pardo, 2010). FRUCTAS has been making demands for years, to no avail.

The recent blockades appeared to have gotten the government's attention at last. David Choquehuanca, Bolivia's foreign minister, blasted Sumitomo, calling it a "transnational that steals our natural resources, plundering tons of minerals every day but does not pay" for the water (AFP 4/18/2010). The San Cristóbal case, however, is only one of many toxic mines across the country. In 2010, the genetics department at the Universidad Mayor de San Andrés in La Paz reported that their tests at the San Jose Mine in Oruro have revealed massive genetic damages to the residents who had inhaled the toxic air near the mine. Scientists reported that 35 percent of mothers and 38 percent of children that they examined had alterations to their DNA as a result (ATB 4/27/10). The lack of government monitoring gives scientists and activists reason to fear that toxic wastes are creating a legacy of health problems that will trouble Bolivia for generations to come.

Morales' government has also given a huge concession to the Indian company Jindal to mine iron ore at the Mutún site near Puerto Suárez on the eastern Amazon area, near the border of Brazil. Activists are worried about the toxic byproducts that seem likely to result from what will be the largest iron ore mine in South America. There are already huge damages from the highway that will support the mine. It is part of the new Bio-Oceanic Highway planned as part of the IIRSA, the South American Regional Infrastructure Initiative, which Denise de Alcantara writes about in this book. The portion of that road already under construction, from Santa Cruz to Puerto Suárez, crosses through the department of Santa Cruz, and across the dry forest lands of the Chiquitano people. The Chiquitano organization, OICH

(Organización Indígena Chiquitana), has been protesting this road since its inception in 2000, and has documented serious damage to the fragile ecosystems along the road's trajectory, including deforestation, contamination of water sources, land conflicts, and loss of flora and fauna. They have also decried the lack of consultation with their authorities, bad working conditions, and the increase in social problems, including prostitution (Erbol, 2010; Bailaba, 2004).

Hydrocarbons: Benefits and Burdens

Of course, Bolivia's main source of income comes from the exploitation of hydrocarbons, mainly in the form of natural gas. Shortly after he was inaugurated, Morales "nationalized" the oil and gas installations, living up to his campaign promises and putting into effect the so-called October agenda — the demands of popular sectors that shut down the country in the 2003 gas war and forced the resignation of then president Sánchez de Lozada. The nationalization turned out to be more of a renegotiation with the transnational corporations who carry out the exploration and exploitation of Bolivia's gas resources. That has proven extremely important for the country, as the government now receives millions of dollars in royalties and taxes, which (along with multilateral debt relief programs) has allowed it to pay off the great majority of its debt, and also begin an ambitious redistribution program.

The 2005 oil and gas law established that the direct taxes from natural gas profits, the IDH (Impuesto Directo de Hidrocarburos), should be distributed to public universities, municipalities, departments (regional governments), and indigenous groups. Royalties are also distributed to the departments and to the central treasury through a complicated formula (see Laserna et al., 2009). At the national level, the government has spent about 30 percent of its IDH monies on a popular retirement account for senior citizens, called the *Renta Dignidad*. Under this program, all Bolivians over the age of 60 now receive about $300 a year. There is also a subsidy for children attending school, called the *Juancito Pinto* Program, in which each child who attends school receives about $25 per year to encourage attendance and minimize dropouts. In 2008, the program was extended to cover all students from first to eighth grades — this means more than 1.7 million students receive the payment each school year.

The IDH also partially funds a $600 million home construction program that provides loans to low-income families. Additionally, with Cuba's help, the central government has funded a national literacy program in Spanish, Quechua, and Aymara; a "zero malnutrition" program to eliminate childhood malnutrition; and a large campaign for eye health, providing check-ups and glasses for those who need them as well as over 100,000 eye surgeries (Weisbrot and Sandoval, 2007). Some question whether these are truly forms of redistribution, since they do not raise the standard of living in a significant way. Indeed, there has been very limited overall social spending by the government and very little improvement in the overall poverty rate in Bolivia since Morales' election (Weisbrot, Ray, and Johnston, 2009:16–19). Nevertheless, these programs are an important beginning and they are enormously popular among Bolivia's poor and indigenous majority, Morales' main constituency.

But what are the effects of oil and gas development on local people? Most of the oil and gas wells — about 350 wells — are located in the Chaco area, in the dry lowlands in the southeast. Eighty-three percent of the reserves lie under the lands of the Guaraní people, who through the land reforms of the 1990s were able to get collective title to small areas of their traditional territories, called TCOS (*territorios comunitarios de origen*). Geographer Thomas Perrault has documented the struggles of one TCO, the Itaka Guasu TCO, where the largest gas field, Margarita, is located. Margarita is operated by a consortium of transnational oil companies, headed by Repsol and Maxus. Perrault documents the dismissive approach the companies took as they planned and carried out the exploration and development of the wells, ignoring Guaraní protests and demands, and polluting the lands of the communities in the process of drilling and extraction. Years of protest by the people and the *Asamblea del Pueblo Guaraní*, the APG, have produced little change, although a 2004 blockade and takeover of the well forced the government to establish a fund by which 2 percent of gas rents paid to the state would be reserved for development projects in Guaraní communities (Perrault, 2008:12). Getting them to actually pay out has proved difficult, however, and the companies continue their pollution unabated. In the weeks prior to the summit, the APG and six communities of Guaranís affected by the ongoing exploitation staged a large march from Yacuiba to Villa Montes to call attention to the continuing problems, like companies failing to consult with them, as is required by ILO Convention 169, to which Bolivia is a signatory, as well as the new Bolivian constitution (see Erbol 2010a, 2010b; El Nacional 2010; and APG 2010).

The Coming Issues: Lithium and Dams

These brief descriptions demonstrate what scholars and activists have been saying for years: Bolivia's historical extractive legacy is ongoing and it continues to create terrible environmental damages, most of which are borne by indigenous peoples at the local level. Two recent projects take this point one step further and reveal that Morales is aggressively pursuing several new megaprojects that have the potential for devastating impacts.

First are the giant new lithium fields in the vast salt flat in southwestern Bolivia, the Salar de Uyuni. Because ultra-light lithium batteries will charge cell-phones and hybrid and electronic cars in the new "green" future, there is a huge and growing demand for lithium. It could challenge petroleum as the dominant fuel of the future (Wright, 2010). The good news for Bolivia is that the newly discovered fields contain about half the world's known lithium — 5.4 million tons — prompting many to predict that Bolivia will become the "Saudi Arabia of lithium" (Howard, 2009, Romero, 2009, Wright, 2010). Morales broke ground on a Bolivian-owned lithium processing plant in 2008, and plans to begin making batteries within a year or so. He is also determined to avoid the history of raw material exploitation that began with colonial silver mining in Potosí. He told journalist Lawrence Wright that the Salar "cannot be another Cerro Rico," referring to the silver mine in Potosí that is the symbol of colonial extraction. Bolivia can "no longer export its raw materials without benefitting from the additional values that the industrialized world gained from them" (Wright, 2010:3). Eventually, with the help of joint ventures — Bolivia is in consultation with companies from Japan and France — Bolivia wants to produce electric cars. What could be more emblematic of consumer-based industrialization than more cars?

But like any other non-renewable resource, producing lithium will take its toll on the fragile ecosystem of the Salar, and the Rio Grande delta where flamingos breed. To exploit the lithium, Bolivia will create large brine beds and evaporation ponds, and then re-inject the left over salt, increasing salinity of the rivers, which local people use to irrigate their farms. The simplest way of processing lithium involves mixing magnesium with the lithium, producing toxic magnesium hydroxide (Meridian Report 2008). Activists and community members are worried about contamination if this method is chosen. Environmentalists are also concerned about the dangers of highly corrosive lithium hydroxide, produced when lithium combines with water. During the rainy season, the Salar often floods — which is why the birds come to breed there. The *Foro Boliviano sobre Medio Ambiente y Desarrollo*, or the Bolivian Environmental Defense League (FOBOMADE), warns that the government has not carried out sufficient water or environmental studies yet, so it is difficult to project what the long-term effect might be (Howard, 2009). After an analysis of the costs and benefits of lithium production in the Bolivian, Argentine, and Chilean deserts (called the Andean Triangle), the independent energy think tank Meridian Institute of France concluded:

> To extract enough lithium to meet even 10 percent of global automotive demand would cause irreversible and widespread damage to these environments [. . .] Liton propulsion is incompatible with the notion of the green car. (Citing Howard, 2009; Meridian, 2008: Section 5.3)

But there are other perspectives. American scholar Nathan Frieberger is studying the communities in the Salar and the engineers working on the lithium project. While there is enormous pressure from the government to produce lithium soon, he suggests that this project has the potential to change the existing model of resource extraction. Precisely because classic-style exploitation will likely damage the unique nature of the Salar, local communities and engineers are pushing to rethink ways of managing the region, trying to incorporate local ways of caring for the land (Personal communication, La Paz Aug. 4, 2011)

The second mega-project looming on the horizon is the Cachuela Esperanza dam planned as part of the IIRSA initiative in the Beni region of the Bolivian Amazon. This dam is part of the Madeira River Hydroelectric Complex, the largest hydroelectric project in the Amazon. The project will dam the second largest river of the Amazon basin, inferior only to the Amazon River itself. The Madeira River carries half of the sediments of the entire basin and feeds one of the most biologically diverse regions of the world that is shared by three countries: Bolivia, Brazil and Peru. The Complex involves the construction of four hydroelectric plants: two in Rondônia, Brazil, near the border with Bolivia, are under construction; a third one, Cachuela Esperanza located in northern Bolivia, is in advanced stage of studies; and the fourth one, Guayará-Mirin located in bi-national (Brazil-Bolivia) waters, is still in initial studies (Bank Information Center 2010).

Indigenous peoples all along the river's trajectory have protested the dams, which will inundate millions of acres of lands and forests, destroying wildlife and ecosystems. A particular concern is the fish population, one of the most diverse on the planet. Critics claim that the dams will disturb mercury deposits — refuse from mining over the years — which will be ingested by the fish, and then by the river dwelling populations who subsist on those fish (Denvir and Riofrancos, 2008). The Morales government initially raised environmental concerns about the project, but has since given its tacit agreement to it. Observers argue that this must be seen in the context of conflict between Brazil and Bolivia over Morales' gas nationalization, which forced the Brazilian gas company Petrobras, one of the largest buyers of Bolivian gas, to pay higher taxes and royalties. Bolivia cannot afford to lose Brazilian income and investment, "so it has toned down its criticism of the Madeira project" (ibid.).

The New Extractivism

How do we make sense of this shocking dissonance between Morales' words about protecting Mother Earth and these images from Bolivia of the environment and indigenous communities under attack? While many simply accuse Morales and the MAS of hypocrisy, I want to offer a few other complicating factors. First, it is unfair to blame Morales for Bolivia's long history of natural resource development. Bolivia has exported primary materials since the Spaniards began mining gold and silver, through the long years of tin, to the recent years of natural gas. As Linda Farthing put it, "Bolivia's history and environment have been dominated by relentless extraction" (Farthing, 2009:25). In essence, Morales took on a country that was already "primary resource-dependent," and it will take some time to alter that model (Denvir and Riofrancos, 2008). Many of the projects described herein — natural gas in the Chaco, the IIRSA roads and dams — are the legacy of previous administrations, especially those of the neoliberal governments seeking to develop Bolivia through regional economic integration. As a result, Brent Kaup (2010) argues that the Morales government continues to be constrained in its options by the "path-dependent" effects of the neoliberal years, which set Bolivia on a development trajectory that is very difficult to change. While it was able to renegotiate the rents from its natural gas resources, it was not able to radically change the material constraints of gas extraction, transport, and use, which keeps Bolivia supplying gas benefitting its neighbors' development projects more than its own (Kaup, 2010).

Second, the Morales government is not just making public performances of *pachamamismo* — it is also taking a leading role in pushing international institutions to rethink their approaches to global climate change. For instance, Morales and his UN representative, Pablo Solón, have relentlessly pushed back against the idea of a market-based model of carbon offsets, arguing that such a model does not resolve the underlying causes of deforestation and degradation of forests. They have been key players in the debates over programs like Reducing Emissions from Deforestation and Forest Degradation — (REDD) and (REDD+) — in which industrialized countries would purchase offsets and establish funds to pay developing countries to maintain forests. Morales and Solón have argued that these "false solutions" only reinforce the capitalist production responsible for climate change in the first place, while attempting to drag developing countries like Bolivia further into the logic of the market.

Lastly, they argue instead for a change in the consumer patterns of the industrialized countries and high taxes on emitting sources in the North to compensate for those who still protect the forests (OBIE/CEDLA 2011). These arguments have ricocheted across the transnational environmental policy sphere, causing a substantial rethinking about the viability of carbon offset programs. Yet, Bolivia must tread carefully here, so as not to endanger its chances of receiving the substantial funds that flow from the REDD programs to forested countries. Bolivia has the sixth largest extension of tropical forests in the world, estimated at between 50 and 60 million hectares (International Tropical Timber Council 2011). This has enormous value in terms of climate change mitigation, and Bolivia has continued to work with the U.N. toward a Plan of Action to prepare it for participation in Reducing Emissions from Deforestation and Forest Degradation (REDD+). This includes passing national legislation to protect forests, training government and non-governmental agents, and working with local indigenous groups, on whose land 80 percent of all forests lie (Ibid).

Nonetheless, the Morales government must meet the urgent needs of the poorest country in South America. As was stated earlier, Morales and his government have used income from natural gas sales for the welfare of the poorest and most marginalized people of Bolivia. Vice President Alvaro García Linera made clear the government's priorities:

> We are going to construct highways, we will drill wells, we will industrialize our country preserving our resources in consultation with the people, but we need resources to generate development, for education, transport, and the health of our people. We are not going to turn ourselves into park-rangers for the powers of the North who live happily, while we continue in poverty. (Erbol 2010c, my translation)

This argument is a perfect example of what Eduardo Gudynas, a researcher at the Latin American Center for Social Ecology in Uruguay, calls "progressive neo-extractivism," which he defines as an emerging national development model of progressive governments like Morales', based upon the appropriation of nature and the export of primary materials. He says that this model differs a little from the traditional extractivist model because "the state plays a more active role, and gives extractivism a greater legitimacy because it redistributes some of the surplus to the population." Nevertheless, "it still repeats the negative environmental and social impacts of the old extractivism" (Gudynas, 2010:1; see also Bebbington, 2009).

Far from repudiating the dependence on international companies and markets, as they propose, these progressive leaders maintain a dependence, replacing the negative connotations of exports and world markets with a new discourse about globalization and competition (Ibid.:4). Defending resource exploitation as part of national sovereignty and necessary for national development, governments reject protests and accuse indigenous groups and peasants of impeding development or generating damages to the country (Ibid.:7, Bebbington, 2009).

When Morales was asked during the Cochabamba Summit about the protests at the San Cristóbal mine, he responded:

> First, that is a concession that is legally guaranteed and armored by the previous governments. It's the legacy of the neoliberal governments. But in addition, the people in the area know that the company has negotiated with them. They've created a foundation to give money to community members and the experience is that such kinds of agreements, blackmail or prebends, are not a solution. Those are not eternal. And those who are culpable are the leaders of the communities who agreed to enter into agreements with the company. There's also a political component. When the right lost in the municipal elections, the next day, they began to wage conflicts. So there's an internal issue there. (DN 4/23/10)

Similarly, when asked about Mesa 18, Morales dismissed their protests as manipulated by non-governmental organizations:

> You want me to tell you the truth about working group 18? That's a business of the NGOs and the foundations. The indigenous brothers and sisters had never before had an indigenous working group within the seventeen. But since it's a question of justifying investments by the NGOs, then they set up working group 18. Now, the internal debate. Those foundations, NGOs, said, "Amazon, no oil". So they're telling me that I should shut down oil wells and gas wells. So what is Bolivia going to live off of? So let's be realistic. But since these foundations and NGOs justify using some of the indigenous brothers and sisters — I don't blame my indigenous brothers and sisters. They use the leaders to justify their good salaries and their own way of life. (Ibid.)

Here, we see all the points Gudynas makes: Morales justifies the model by a pragmatic obligation to support the country –"*seamos realistas*" — and dismisses any objections to it, here with a classic Morales blaming of outsiders (see Postero, 2010). So, Gudynas' argument remains compelling and quite accurate in the case of Bolivia, but ends up being more descriptive than explanatory. The question it leaves unanswered is *why* these progressive governments take this tack. Bebbington and Humphreys (2010) remind us that the push towards extractives is not only a matter of presidential fiat but is also an artifact of national and global political economy. Here we return to the path-dependent effects of Bolivia's extraction history and the role that history plays in the Bolivian national imaginary.

The Inherent Tensions in the MAS Agenda

The main reason why Morales pushes natural gas and all the mega projects is because it is a fundamental part of his mandate from his constituents. Morales came to power with a mandate to overcome neoliberalism, through which the *patrimonio* of the country was given away to transnational corporations. This was the clear message of the water and gas wars of 2000 and 2003, and the reason President Morales "nationalized" the hydrocarbon sector first thing. This is a matter of great importance to most Bolivians, who felt the neoliberal leaders of the 1990s cared more about putting dollars into their own pockets and those of their international corporate friends than building a just and prosperous Bolivia. This was also a call to regain sovereignty over natural resources and to redistribute the profits from them to the rightful owners, the Bolivian people. Note that this is not just a call to keep the profits at home, but also a call to industrialize at home. Bolivians understand that in a global economy they cannot only provide raw materials, but must also provide added value. So, the majority of Bolivians, including many indigenous people, are proud of the nationalization, and delighted that Morales and his associates reversed the unfair terms of the gas business. They *want* lithium to be developed, and they want their standards of living to improve. This is part of *pachakuti,* the turning of the timetable, the change of destiny. This is the time for the formerly poor to receive their fair share.

But Morales also came in with a second, and equally important agenda: to decolonize society, to recognize the indigenous peoples whose lands and cultures form the basis of Bolivia, and to re-incorporate their rights and values into the society. This is the fundamental tension that Morales faces as he articulates the indigenous "Save Mother Earth" discourse while also plunging ahead with the mega-projects described earlier in this chapter. This tension is inscribed in the new Bolivian constitution, drafted in a popular constituent assembly and passed by the Bolivian people in a referendum in December 2009. Article Eight describes the underlying moral and ethical values of the state:

> The State assumes and promotes as ethical-moral principles of the plural society: ama qhilla, ama llulla, ama suwa (*no seas flojo, no seas mentiroso ni seas ladron*) [this is an Aymara moral code, don´t be lazy, don´t lie, don´t steal]; suma qamaña (*vivir bien*) [an Aymara ideal of the good life], ñandereko (*vida armoniosa*) [a Guaraní ideal of harmonious life]; teko kavi [a Guaraní admonition to live well and wisely]; ivi marei (*tierra sin mal*) [a Guaraní idealized notion of the land without evil]; and qhapaj ñan (*camino o vida noble*) [a Quechua call to follow the noble path]" (Art. 8, § 1)

Part II explains the goals of the state:

> The State is sustained in the values of unity, equality, inclusion, dignity, liberty, solidarity, reciprocity, respect, complementarity, harmony, transparency, equilibrium, equality of opportunity, social and gender equity in participation, common wellbeing, responsibility, social justice, distribution and redistribution of products and social goods, to live well.

This is what Morales was talking about at the summit. In the next constitutional article, we see that one of the state's fundamental functions is to push development of natural resources and their industrialization:

> The ends and essential functions of the state include: 6. To promote and guarantee the responsible and planned development (*aprovechamiento*) of natural resources, and ***to impel their industrialization***, through development and the strengthening of the productive basis in their different dimensions and levels, along with the conservation of the environments, for the wellbeing of present and future generations.

Other articles give indigenous people the right to control their territories, to exercise self government/autonomy over their communities, and, most relevant to this discussion, the right to participate in the benefits from the exploitation of natural resources in their territories, and to be consulted in good faith by the government prior to any exploitation in their territories (Article 30).

So, despite the fact that Morales espouses an indigenous rhetoric at the international level that purports to provide solutions, he and Bolivia are facing the same dilemma everyone else is: how to develop *and* live well, how to find a path to sustainable industrialization *and* to share the benefits and burdens fairly and justly.

Obviously, Bolivia has a long way to go to achieve the "equilibrium," "harmony," and "social justice" the constitution calls for. To accomplish it, Morales and his government must overcome centuries-old economic patterns and deal with Bolivia's vulnerable geopolitical position in a regional and global market, while keeping its constituents at home satisfied. Ironically, rents from gas have made all of that much easier — as it has given Morales freedom from international aid and allowed the kinds of redistribution discussed earlier in this chapter– but also more difficult — as lowland elites have pushed harder for their share of the royalties through the contentious autonomy movement (Gustafson, 2006; Fabricant, 2009). Now the government is finding that there are even greater costs to this development model: indigenous people who have borne the burden of these policies are no longer going along with the MAS agenda but are putting the contradictions in Morales' face. Morales must either sacrifice these victims (Gudynas, 2010) or make some big changes.

Ruptures at Pachamama's House

It has become increasingly clear that the Morales government is opting for the sacrifices. In the process, the fictitious unity of the "indigenous people of Bolivia" has begun to unravel in public, revealing what anthropologists have long known: that indigeneity is a constructed concept, the meanings of which are forged in particular relational fields of force. Morales and the MAS have done a masterful job at claiming indigeneity as a discursive backdrop for their political and economic agenda. Yet this is based on a particular idealized vision of indigeneity, which posits the "indigenous people of Bolivia" as a coherent subject that can be known and represented by him. This does not accurately reflect the realities of Bolivia, which has over thirty different groups of people who claim indigenous status, living in very different conditions, with different histories, languages, and customs.

Despite their differences, most of these groups threw their fortunes in with the MAS, taking part in the process of change he represented. Yet this was a fragile alliance built more than anything on the desire to defeat a common enemy, the white-*mestizo* political class that had held power since the founding of the republic. That alliance held over the first five-year Morales administration, because the Right wing opposition forces headed by the lowland *camba* elite threatened the promised changes. Now that their threat is greatly reduced, and the new constitution has restructured the state, that alliance is breaking down. The need to hold onto the fiction of indigenous unity is dissolving.

The protests of Mesa 18 in spring 2010 were a sign of the Morales government's willingness to disregard indigenous people's concerns about their land and environments. Then, during the congressional debates in 2010 over the law implementing the new constitution's notions of "indigenous autonomy", the ruptures became even more obvious. When the Confederation of Indigenous Peoples of Bolivia (CIDOB), the lowland indigenous organization, protested that the MAS version of the law did not take into account their demands (regarding congressional representation, procedures for community buy in, and the need for territorial protections, among other things), the government began a campaign to discredit them.

First, Morales disparaged them as being racist, claiming their demands for more dedicated seats in the Congress denied other Bolivians equal representation. Then, he argued that the indigenous groups were being greedy trying to protect their territories from needy colonizers. Morales met with campesino leaders proposing to change the agrarian reform laws to eliminate indigenous territories, which, they said, favored small groups while peasant farmers were limited to the worst lands. Morales said he agreed with the proposal because it would terminate this form of *latifundismo* (Bolpress, 2010). Finally, he brought in the big guns: he accused CIDOB of being financed by the United States Agency for International Development (USAID), implying that they were being manipulated by the American interests. Eventually, Morales backed down, and made some concessions, including several CIDOB demands in the final Autonomy Law. However, most observers agree the new law provides for a much weaker form of self government than indigenous groups demanded at the Constituent Assembly. The debate and the government's responses to the march left bitter feelings among lowland groups, many of whom are beginning to form alliances with the lowland elite again.

At the time this chapter goes to press, August 2011, lowlands indigenous peoples joined in another march to protest government plans to put a highway through the fragile Amazon region called the TIPNIS (*Territorio Indígena Parque Nacional Isiboro Sécure*), which is both an indigenous territory and a national park. Once again the government has accused CIDOB of being manipulated by USAID (El Deber 2011). Once again, the fate of Bolivia's Amazon forests and communities is at the center of Bolivian politics.

Conclusion

While President Morales has been adept at utilizing indigenous culture and rhetoric to project an international image of environmental savior, the realities at home in Bolivia are far more complex. On the one hand, as part of his commitment to decolonizing the country, Morales and his government put forward alternatives to the colonial-modern capitalist logic that continues to push the earth to its destruction. To do so, they have adapted the cultural notion of *vivir bien*, or living well, from its historical meanings in various indigenous languages and local contexts into a national political economic strategy articulated in the constitution and put forward in international spheres. On the other hand, responding to the particular history of Bolivia's engagement with capitalism and especially neoliberal forms of it, Morales and his government must make the economy benefit those previously marginalized, and they must do so quickly. Moreover, this fundamental tension takes place within a complex field of forces where global capitalism constrains the possibilities.

What, then, does *vivir bien* mean in this context? Until now, indigenous peoples harmed by the extractions have often been sacrificed to the "greater good" of industrializing the country. In recent years, however, they have begun fighting back with increased vigor. Yet, *vivir bien* is not just a token rhetorical tool. Rather, it has sufficient moral and cultural significance such that its insertion into the debate has changed the discursive and material field in which contestations over resources and distribution are occurring. It is a new "language of contention" as William Roseberry (1996) would put it, a standard against which all actors, including the government, will be held to account.

Note

1 Bebbington and Humphreys Bebbington report that Ecuadoran indigenous leaders also identified with the movie, although some reflected on the common critique heard in the United States as well, that the movie's hero was a white man, reinforcing the idea that indigenous need salvation from outside their communities. (Bebbington and Humphreys; Bebbington, 2010)

Bibliography

AFP, 2010. Bolivia slams Japan mining firm for 'plundering' resources.

APG (Asamblea del Pueblo Guaraní), 2010. Indígenas Guaraní del Departamento de Tarija Marchan Exigiendo al Gobierno Se Respete el Derecho a la Consulta. http://www.cidob-bo.org.

ATB, 2010. Estudio de la UMSA revela que aire tóxico de mina San José afectaría a genes de pobladores. http://www.youtube.com/watch?v=nAXYfOOGt_4&feature=player_ embedded

Bailaba, Jose, 2004. Interview with Chiquitano leader, INDUCCIÓN Y ETNOCIDIO CHIQUITANO — II, in Consignas. http://consignas.espacioblog.com/post/2005/09/22/induccion-y-etnocidio-chiquitano-ii.

Bank Information Center, 2010. Río Madeira Complex, http://www.bicusa.org/en/Project.10138.aspx

Bebbington, Anthony. 2010. The New Extraction: Rewriting Political Ecology in the Andes? *NACLA Report on the Americas* 42(5): 12–15.

Bebbington, Anthony and Humphreys Bebbington, 2010. An Andean Avatar: Post-neoliberal and Neoliberal Strategies for Promoting Extractive Industries. BWPI Working Paper 117, (forthcoming in *New Political Economy*).

Bolpress 2010. El gobierno y la CSUTCB planean recortar las TCOs indígenas, August 4, 2011. http://www.bolpress.com/art.php?Cod=2011080401

Buenos Aires Herald, 2010. Evo Morales 'identifies' with Avatar film, January 12, 2010.

Cooper, Elizabeth. 2010. Bolivia's Climate Summit: "The People's Agreement." The Democracy Center, http://democracyctr.org/blog/

Carvajal, Rolando, 2010. Sumitomo confirma que tributó 358 millones en 10 años, pero nada por el agua que utiliza. http:/boliviaenvideos.com/2010/04/sumitomo-confirma-que-tributo-358.html

Democracy Now, 2009. Bolivian President Evo Morales on Climate, Copenhagen, and Capitalism. December 17, 2009. http://www.democracynow.org/2010/12/17/ bolivian_president.

Democracy Now, 2010. Bolivian President Evo Morales on President Obama: "I Can't Believe a Black President Can Hold So Much Vengeance Against an Indian President." April 23,2010. http://www.democracynow.org/2010/4/23/bolivian_president.

Democracy Now, 2010. Why is the US Cutting Off Climate Aid to the Poorest Country in South America? — Bolivian Climate Negotiator Angélica Navarro, April 21, 2010. http://www.democracynow.org/2010/4/21/why_is_the_united_states . . .

Denvir, Daniel, and Thea Riofrancos, 2008. How Green is the Latin American Left? A Look at Ecuador, Venezuela, and Bolivia. Upside Down World, April 2008. http://upsidedownworld.org/main/content/view/1203/1

El Deber 2011. Bolivia sospecha que EEUU alienta la marcha indígena, August 21, 2011. http://www.prensaescrita.com/adiario.php?codigo=AME&pagina=http://www.eldeber.com.bo

El Nacional, 2010. Indígenas del Chaco cerrarán el Parque Aguaague a las empresas petroleras. Tarija, Bolivia, 4/18/10.

Erbol, 2010a. Indígenas marcharán el lunes contra el gobierno por contaminación petrolera en el Gran Chaco.

Erbol, 2010 b. Indígenas marchan contra el gobierno y demandan consulta sobre hidrocarburos. April 14, 2010.

Erbol, 2010b García Linera: No nos vamos a convertir en guardabosques del norte. La Paz, Bolivia, April 26, 2010.

Fabricant, Nicole, 2009. Performative Politics, The Camba Countermovement in Eastern Bolivia. *American Ethnologist* 36:4: 768–783.

Farthing, Linda, 2009. Bolivia's Development Dilemma: Development Confronts the Legacy of Extraction, *NACLA Report on the Americas* 42(5): 25–29.

Gudynas, Eduardo. 2010. The New Extractivism of the 21st Century, Ten Urgent Theses about Extractivism in Relations to Current South American Progressivism. Americas Program Report (Washington, DC: Center for International Policy, January 21, 2010).

Gustafson, Bret. 2006. Spectacles of Autonomy and Crisis: Or, What Bulls and Beauty Queens have to do with Regionalism in Eastern Bolivia. *Journal of Latin American Anthropology* 11: 2: 351–379.

Haglund, Christina. 2008. A River Turns Black, Enron and Shell Spread Destruction across Bolivia's Highlands. In *Dignity and Defiance, Stories from Bolivia's Challenge to Globalization*, Jim Shultz and Melissa Crane Draper, eds. Berkeley: University of California Press.

Howard, April, 2009. Salt of the Earth, Bolivia extracts its lithium, environment be damned. *In These Times*, May 6, 2009.

International Tropical Timber Council 2011. Status of Tropical Forest Management 2011. www.itto.int/direct/topics/topics_pdf . . . /topics_id=2648&no=2&disp=inline

López Pardo, Claudia. 2010. Mining protests overshadow climate summit, *The Guardian*, 7/05/2010, http://www.minesandcommunities.org/article.php?a=10093

Kaup, Brent. 2010. A Neoliberal Nationalization? The Constraints on Natural-Gas-Led Development in Bolivia. *Latin American Perspectives* 37(3): 123–138.

Klein, Naomi, 2010. A New Climate Movement in Bolivia. *The Nation*, April 21, 2010.

La Razon, 2009. Pando: el Gobierno fuerza la colonización con gente afín al MAS, 8/11/2009.

Laserna Rojas, Roberto, Carlos Miranda Pacheco, and Mario Napoleón Torrico. 2009. Control estatal de recursos naturales y rentismo, in *Poder y Cambio en Bolivia, 2003–2007*, La Paz, Bolivia: Fundación PIEB. (pp. 9–60)

Meridian International Research, 2008. The Trouble with Lithium 2: Under the Microscope, William Tahil, Research Director.

Morales Ayma, Evo, 2009. Address by H.E. Mr. Evo Morales Ayma, the President of the Plurinational state of Bolivia, General Discussion on the Occasion of the 64th Session of the General Assembly of the United Nations.

Munkton, Stewart. 2009. Morales named 'World Hero of Mother Earth'. *Green Left Weekly*, http://www.greenleft.org.au/node/42396.

OBIE/CEDLA, 2011. Bolivia y el Reto de No Ceder en la Defensa de la Pachamama ni Perder el Dinero del Carbono, El Observador, Observatorio Boliviano de Industrias Extractivas/Centro de Estudios para el Desarollo Laboral y Agraria. February 2011. La Paz, Bolivia.

Perrault, Thomas. 2008. Natural Gas, Indigenous Mobilization, and the Bolivian State. Identities, Conflict and Cohesions Programme Paper Number 12, United Nations Research Institute for Social Development.

Postero, Nancy. 2010. Bolivia's MAS Government, Building Indigenous Popular Hegemony in Bolivia. *Latin American Perspectives* 37(3):18–34.

Romero, Simon. 2009. In Bolivia, Untapped Bounty Meets Nationalism. *The New York Times*, February 3

Roseberry, William. 1996. Hegemony, Power, and the Language of Contention. In *The Politics of Difference*, Edwin Wilmsen and Patrick McAllister, eds. Chicago: University of Chicago Press.

Rosenthal, Elisabeth. 2009. In Bolivia, Water and Ice Tell a Story of a Changing Climate. *New York Times*, 12/14/2009.

Weisbrot, Mark and Luis Sandoval. 2007. Bolivia's Economy- an Update. Washington, D.C.: Center for Economic Policy Research Issue Brief. http://www.cepr.net/index.php/publications/reports/bolivia-s-economy-an-update. Accessed September 16, 2008.

Weisbrot, Mark, Rebecca Ray, and Jake Johnston. 2009. The Economy During the Morales Administration. Washington, D.C.: Center for Economic Policy Research Issue Brief. http://www.cepr.net/documents/publications/bolivia-2009-12.pdf. Accessed January 27, 2011.

World People's Agreement, 2010. World People's Conference on Climate Change and Mother Earth, April 2010, Cochabamba, Bolivia. Text at http://www.cmpcc.org.bo/PEOPLES-AGREEMENT.

Wright, Lawrence. 2010. Lithium Dreams; Can Bolivia become the Saudi Arabia of the electric car era? *The New Yorker*, March 22, 2010.

Vidal, John, 2009. Evo Morales stuns Copenhagen with demand to limit temperature rise to 1C. *The Guardian*, December 16, 2009. http://www.guardian.co.uk/environment/2009/dec/16/Morales-morales.

7 IIRSA: Social and Environmental Challenges on a Strategic Development Effort

Denise de Alcantara Pereira

Introduction

The focus here is on some of the regional development efforts done during the 1970s and 1980s to integrate remote and underdeveloped areas of the country, and relating them to what is being proposed by IIRSA (Initiative for the Integration of Regional Infrastructure in South America). Brazil is the largest South American country, both in terms of its territory and in its population size, and in the fact that it holds the largest portion, 60 percent, of the Amazonian rainforest. Recent studies and research[1] show that Brazil has reached a position of economic and political stability; its importance and influence in the region has grown.

Over the past years, we have seen efforts to integrate South America through the development of infrastructure — mainly telecommunications, electricity and transportation — that pose a whole range of questions. One of the most important questions relates to the procedures that are being used to monitor and control the major work done and how projects are being implemented, considering the fast pace with which the public and private sectors are engaged in mega projects and infrastructure building. Most of the infrastructural and technological projects rely on roads that cut and will cut through untouched natural forests and land, and sometimes will encroach upon remote indigenous communities that have not been previously encountered. As previous experiences in Northern Brazil show, the building of roads is one of the most damaging human actions taken on environmentally sensitive natural forests and landscapes, especially in the Amazon.

However, it is also important to consider that IIRSA is a strategic project for all South American countries, in terms of how it will affect trade, market opportunities, and the export of natural and industrialized products. As a Brazilian architect and urbanist, I usually address micro scale portions of a territory — the built environment, man-place interactions, and the study of urban areas. This kind of perspective can be adjusted to a macro regional perspective — the Amazon basin — because man-environment interactions can be projected into larger environmental settings. It leads to the recognition of a system, of a regional network connecting places and people and, asks the question about accessibility and transportation systems concerning people, products, and technologies. In this sense, other important questions emerge: what kind of transportation systems should be implemented in the Amazon? Are the ones being currently implemented aware of sustainability issues and protection of Amazonian ecosystems and biodiversity?

To begin answering these questions, this study looks back to concepts like *dependency theory*,[2] *consensual hegemony*,[3] *development hubs, sustainable development* and *democratic institutions*. Such concepts help us understand how far IIRSA, in conjunction with South American governments, plan to develop each affected country through a scheme of regional integration, rather than acting as independent states. *Dependency theory* may explain "a situation in which the economy of certain countries is conditioned by the development and expansion of another economy to which the former is subjected; some countries (the dominant ones) can expand and be self-sustaining, while other countries (the dependent ones) can do this only as a reflection of that expansion".[4] This was the case until countries like Brazil and Venezuela became less dependent on financial institutions and on wealthier developed nations that, until recently, had ruled most of the South American countries through economic and political impositions.

In the consensual hegemony proposition, we find arguments to understand Brazil's position in relation to other South American countries, where the intention was to induce a desire of mutual benefits that could subtly embed (rather than impose) Brazilian interests, aspirations and strategies in the region.[5] IIRSA's transnational character makes the consensual hegemony proposition a fine tool to achieve a balanced and equitable negotiation ground for all countries involved. 'Development hubs' is perhaps the most important methodological tool regarding IIRSA's strategic approach, as it attempts to address the economic, social, and environmental dimensions of physical integration and their interplay in the wake of a project's development process. All these concepts are necessary to understand the historical political process that has shifted over the years in relation to Brazil's global position, and Brazil's renewed attempts at regional integration.

South American Biomes

Distinct biomes are visible in South America and each has its own characteristics and sensitiveness. According to Christopher Brown,[6] it is relatively easy to define the major specific biomes from a macro scale point of view, but the main challenge is defining their boundaries. There are no precisely defined borderlines and in most cases they overlap. This makes it almost impossible to establish political and environmental regulations in areas of intermingling biomes.

During a road trip I took in 1990, from Rio de Janeiro to Mato Grosso do Sul through São Paulo, I noticed the drastic landscape changes in the border limits of the states of São Paulo and Mato Grosso do Sul. When crossing the border between the two states, the neat and tidy landscape with agriculture and pastures of São Paulo's wealthy rural groups, turns into the almost untouched *cerrados*[7] of Mato Grosso do Sul. Twenty years later, this landscape has changed. With the expansion of agro-industry in the center-western states, the *cerrado* biome is now mostly a large scale soy bean monoculture.[8]

Brazil holds about 30 percent of the world's tropical forests. The Amazon is the largest Brazilian continental biome and occupies 4.196,943 square kilometers or 49.29 percent (almost half) of the national territory. The Brazilian Amazon eco-region covers the states of Amazonas, Pará, Amapá, Roraima and Acre and, partially, the states of Maranhão, Mato Grosso, Tocantins and Rondônia. The hydrographic basin of this vast area contains important rivers and tributaries, the most important being the Amazonas River, but there are others like Negro, Tapajós, and Madeira that also carry massive amounts of water. The Amazonas River is the river with the largest amount of water in the world; it originates in the Andean mountains and together with its many tributaries in other countries, it enters Brazilian territory. These countries are Bolivia, Colombia, Ecuador, Guiana, French Guiana, Peru, Suriname and Venezuela.

The Amazonian biome, especially in its southernmost part below the right margins of the Solimões River, concentrates the best lands for agriculture (soy, corn, wheat) and cattle production on large farms. Currently these products travel mostly on roads to reach the closest export ports (Belém and Manaus) and all the way to the southern states of the country. Ports are located on the Atlantic coast; Brazil has no direct access to the Pacific coast. Ships with South American flags have the right to navigate though the international waters of the Amazon rivers, a possibility Brazil does not have because the water flows east from the Andes to the Atlantic. This geographical navigational system is one of the most important and sensitive aspects of IIRSA, since the roads are built to access the Pacific Ocean. According to the plan, to achieve this goal Brazil is investing in opening roads that run through neighboring lands and forests, often still untouched. The ultimate goal of this vast road system is to access the Asian market.

There have been many alternative studies and proposals to find ways to access the Pacific coast. Enrique Amayo's proposal, inspired by one of the most important Brazilian writers of the nineteenth century — Euclides da Cunha — suggests that the best connection between Brazil and the Pacific would be through Peru. He proposes to use hydroways (the numerous existing navigable rivers on both sides of the Andean mountains), the creation of a multimodal system and the construction of railroads over existing narrow roads and paths, in addition to using ship routes through Northern Peru's navigable rivers, instead of opening new roads running through environmentally sensitive forests and natural landscapes. If implemented — he suggests — this proposal would have the least environmental impact in the Amazon basin: [9]

> The best points for Brazil to reach the Pacific Ocean lay in this area. Brazil is a major soy bean producer, and its largest soy crops are located closer to Peru than to the Atlantic coast [. . .]. Peru would benefit mainly from the port construction and corresponding revenues from a port that probably would be the largest on the Pacific Coast of South America.[10]

In his proposal, Amayo also considers border issues related with Colombia, for political and security reasons. With other neighboring countries there are no established borders. Unfortunately, this is no longer an option for the Brazilian government. An agreement has already been signed between Peru and Brazil by which a highway, known as *Corredor Transoceánico Rio Branco–Iñapari*, is already being built through the border of Acre, crossing the Andes, and reaching a port in southern Peru.

Integration and Development Attempts

The Brazilian government has promoted various development strategies for geopolitical and economic reasons. Since the military dictatorship in the 1960s and 1970s, the main goal has been to protect the Amazon region, as a matter of sovereignty, not to protect the environment. The military government shielded the region from foreign speculation and from the richer nations' potential invasive intentions. Different projects to develop and integrate Brazil's Amazon basin to the other regions in the country were set forward. The government's motto was "*integrar para não entregar*" (to integrate to not give it away). For the military, this was an important issue encompassing the idea of bringing people from other regions within Brazil to populate the Amazon basin. The military government ignored the existing indigenous groups who were already occupying these lands and, initially, did not grant any considerations or privileges to them. Back then, the Amazon basin was considered a nonproductive area in need of development at any cost.

The *Operação Amazônia* Project (Amazonia Operation Project) came into being in 1966–67 to integrate and occupy Brazil's Amazon basin through road construction, agribusinesses and industrialization. The *Grande Carajás* Project (PGC) was implemented from 1981 to 1990 with the objective of converting virgin rainforest (11 percent of Brazil's national territory) into an industrial and agro-livestock heartland. The National Integration Program (PIN — Programa Nacional de Integração) provided funds for the construction of 15,000 kms. of roads (the Transamazon and the Cuiabá-Santarém, among others). One of the purposes of this expanding road network was to make land available for new settlements within a 20 kilometers band on either side of the roads. Fiscal incentives to investors in agriculture and industry were put in place.[11]

The projects focused on road construction and the promotion of migration into the Amazon basin. The government offered help in the form of infrastructure, financial benefits and fiscal exemptions for these new settlers and investors. Many settlers responded to the generous governmental offers and moved from other parts of the country, mainly from the South, to establish themselves in different areas of the states of Pará and Rondônia. Most government sponsored projects were very difficult to monitor. Today's monitoring systems like SIVAM[12] did not exist until 2002 and up until then, there were no means of regulating development initiatives in the region. Many projects subsidized at great cost did not provide the promised benefits. To the contrary, large ranches consolidated, replicating the top-heavy land distribution patterns in Brazil with the concomitant displacement of recent migrants. In the late 1980s, fiscal constraints and problems related to international and domestic political reactions to the devastation of the rain forest forced Brazil to deactivate its rural settlement and major development projects in the Amazon basin.

Ever since the region was first occupied by European explorers, until very recently, transportation in the Amazon basin had been by water and air. There were no roads, paved or unpaved, and accessibility was difficult. Travelling from one village to another on boats could take days or weeks. Human settlements were mostly located close to river banks, leaving the forest untouched and preserved. The only ones with access to and knowledge of the forests were the indigenous peoples who lived in a respectful and interactive relationship with nature. Starting in the 1960s until the 1980s, when integration and regional development strategies for the Amazon basin

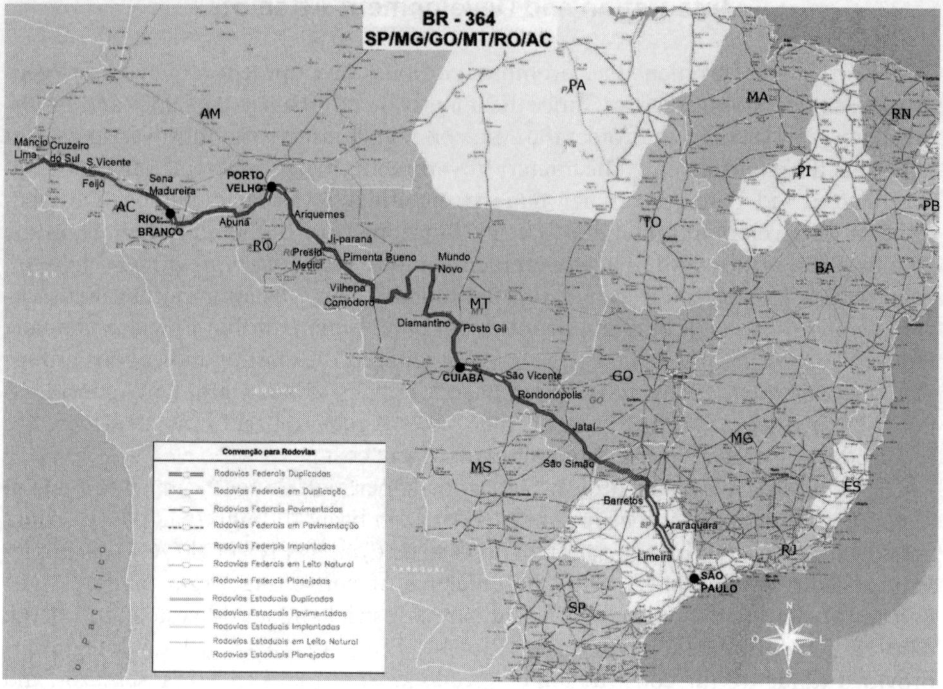

Figure 7.1 Maps showing the road BR-364 and the amplification of the State of Rondonia, where most of the deforestation happened since the first attempts to develop and integrate the Amazon in the 1960s. (*Source*: <http://www2.transportes.gov. br/Site% 20BIT/02.1-Top%20Rodovi%C3%A1rio/03-Loc%20Rodo/localiza%C3%A7%C3%A3o%2 0das% 20rodovias/364.htm>)

began accelerating, three different roads were built. The first major road built to integrate the Amazon basin to the rest of the country was BR-364 (Cuiabá-Porto Velho Road), completed in the 1960s. It was meant to serve as an access route for infrastructural development between Cuiabá and Porto Vellho, but it actually connected a longer stretch from São Paulo to the western border of Acre. The main section of the road crossed the state of Rondônia that began developing in the 1960s (Figure 7.1). Satellite images in the internet[13] show the fishbone pattern created with the construction of BR-364 and the environmental impact it had. The highway BR-364 became a major transportation route for immigrant farmers seeking income producing opportunities in the state of Rondônia. To this day, most of the roads off the BR-364 remain unpaved. Therefore, the devastation caused by the opening of the main and secondary roads — formally or informally, legally or illegally — has expanded dramatically. Deforestation continued at a high speed reaching its second peak in 2004 (17 percent of the total deforestation in the Amazon).[14]

Transamazônia (BR-230) is another major federal road built by the military government in the 1970s. It connects the Brazilian eastern coast in João Pessoa, in the state of Paraíba, to the western border of the country, in Benjamin Constant, in the state of Amazonas and, beyond Brazil's border, to Peru and Ecuador. Planned as

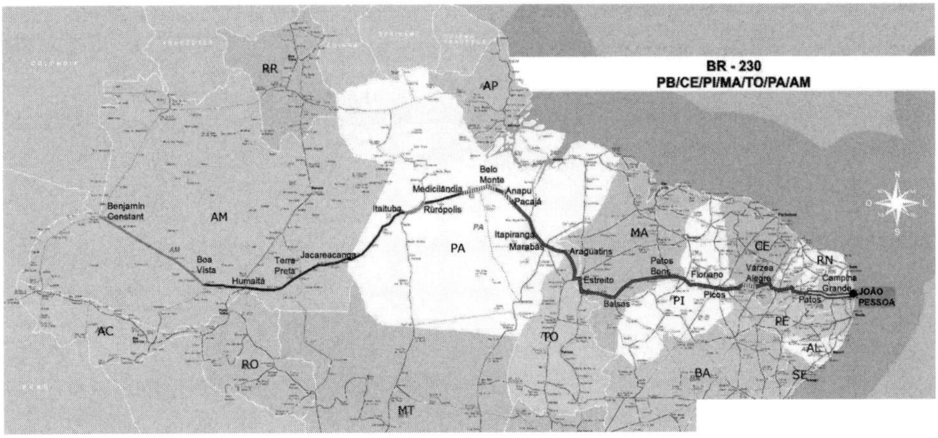

Figure 7.2 Map of Northern Brazil showing the path of the Transamazônica road (BR-230) with the eastern stretch paved and the western one unpaved. (*Source*: <http://www2.transportes.gov.br/Site%20BIT/02.1-Top%20Rodovi%C3%A1rio/03-Loc%20Rodo/localiza%C3%A7%C3%A3o%20das%20rodovias/230.htm>)

a 4,000 kms. road to transversally connect northern and northeastern Brazilian territory, its main purpose was the same as of BR-364, namely, to integrate, protect and bring people to the *undeveloped, unproductive and empty*[15] Amazon (Figure 7.2). Transamazônia was one of the pharaonic projects implemented by the military government during the Brazilian miracle period, when major investments went into mega constructions with international funding. The Rio-Niterói Bridge, the nuclear power plant in Angra dos Reis, and the Itaipu bi-national hydroelectric power plant were some other of these huge projects built during this time period. As of 2011, most of the Transamazônia is unpaved, especially in the "Legal Amazon".[16]

These roads are inaccessible for most of the year because heavy rains make them too muddy and slippery for trucks and cars. Nevertheless, the fishbone pattern has also developed in the wake of the construction of BR-230 and reminds us of how potentially damaging for the natural environment the opening of roads can be. Even in its worst condition, the Transamazônica has triggered 50 kms. inland destruction to each side of the road. Vast stretches of land have been cleared for low-intensity cattle pasture and short-term subsistence agriculture.

Our third case is the Cuiabá-Santarém road (BR-163), most of which is still unpaved (Figures 7.3 and 7.4). Other roads were opened in the 1970s and 1980s, but BR-364, BR-230 and BR-163 are the most impressive ones for their length and also for their almost straight paths. The first one is southwest-northwest in a diagonal pattern, the second one is an east-west horizontal road, and the third one stretches south-north in a vertical line. They are the three major roads that cross the Amazon basin not only providing access to the region, but also causing the destruction of the environment.

As shown in Figures 7.3 and 7.4, road construction in the Amazon basin along BR-163 has involved economic, social, and environmental consequences, such as the fragmentation of habitats, the alteration of microclimates, the introduction of invasive species, the creation of barriers for the movement of wildlife, and the killing of animals

that cross those roads or are simply hunted down. According to Van Dijck "[r]oads play a key role in economic development, yet are among the major driving forces behind the destruction of ecosystems"[17]. Deforestation is concentrated to a 75 percent within fifty kilometers along the main roads, not to mention the side roads that have been opened illegally by "madeireiros" in search for wood (Figure 7.5).

In order to exploit "undeveloped" regions, Brazilian policies were geared towards establishing new settlements in remote areas, a process by which indigenous peoples lost their land, their liberty, and their very lives. In such areas, the landscape was forever changed; roads were built disregarding local topography and hydrology. Entire ecological systems were disrupted, replacing the complex with the simple, eliminating species of plants and animals. Although nowadays we find a new discourse surrounding IIRSA's declarations and actions and a more comprehensive approach to projects that prioritize environmental protection, and that we witness a high level of responsiveness and awareness by the people of the region, we cannot forget the

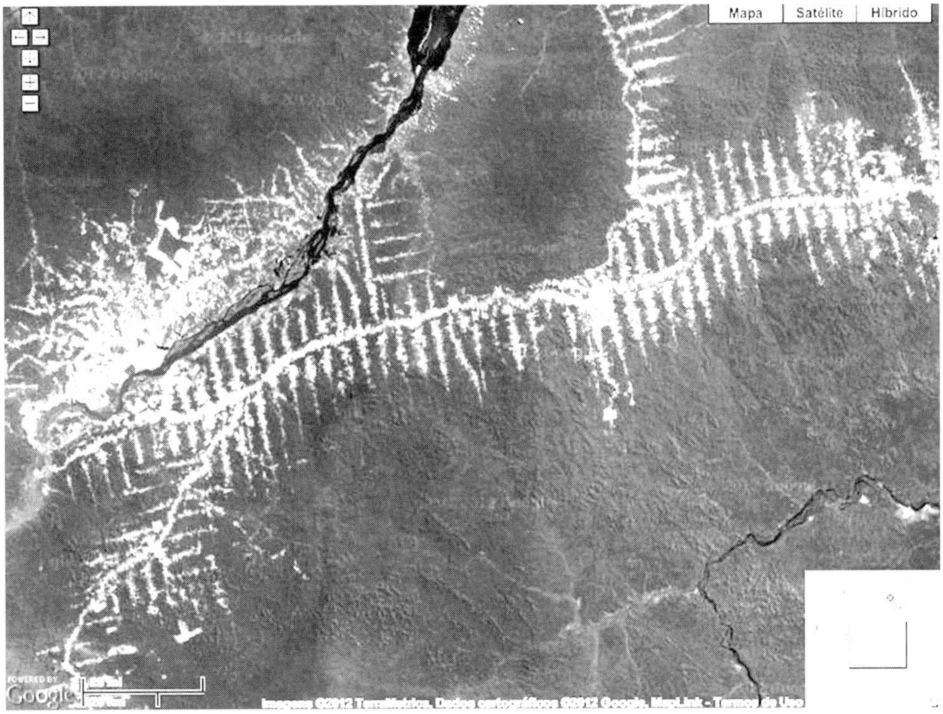

Figures 7.3 (opposite) and 7.4 Map of the BR-163 road and satellite image showing the human occupation along the road. (*Sources*: <http://www2.transportes.gov.br/Site%20BIT/02.1-op%20Rodovi%C3%A1rio/3Loc%20Rodo/localiza%C3%A7%C3%A3o%20das%20rodovias/br-163/gbr-163.htm > and Google Earth satellite image)

Figure 7.5 Large stretches of forest devastated along the main road to create pasture and agriculture on land that is usually inadequate for this purpose. (*Source*: Google Earth)

aggressive and intensive pace with which the projects are moving forward. Maybe at this point there is not enough time left to engage in the necessary and fundamental environmental impact studies and evaluations to comply with the principle of sustainability. However, the costs of not doing it are very high and as of yet unforeseen.

Property titles in Brazil are another tangled and problematic issue, especially in the Amazon basin. A clear land tenure pattern not only helps protect the environment but also entitles citizens to obtain financial support from banks in order to sustainably develop their property. Since the 1970s, the Brazilian Ministry of the Interior has attempted to better regulate property titles along BR-163. The small village of Novo Progresso, in the state of Pará, was chosen as a pilot project for this endeavor because of its location almost halfway between the two extremes of this longitudinal road. Novo Progresso is also the municipality that accounts for the largest tracts of untitled land. The Ministry of the Interior garnered large coverage in the written and televised media. In an interview with a São Paulo newspaper, a pioneer settler defended this government action: "With property titles landowners will work harder to maintain the forest. [. . .] The ones who have titles live off the land, have an address and can be fined or go to jail, and are the ones who worry about not destroying the environment."[18] The newspaper also states that the lack of land titles carries several difficulties, including the inaccessibility to credit from financial institutions and government programs. INCRA (National Institute for Land Reform) was willing to solve the irregular possession of land as it was estimated that each five hectares of land in the Amazon basin was not backed up by any documents or titles. It is land occupied by squatters, the so-called *grileiros*. INCRA attempted to reward the legalization of property holdings with infrastructure and a quick land titling process in order to foment good agricultural practices. Still today (2011) there is a long way to go. Up to now, there remain 104 million hectares (three times the State of São Paulo) of public land with no official allocation.

Importance of IIRSA to Brazil

Brazil's government has been proactively pursuing regional integration in South America. It has actively participated in most of the international agreements with distinct objectives. The Free Trade Area of South America (ALCSA), and the Regional Trade Agreement (Southern Cone countries, MERCOSUL) were the first agreements signed to initiate a process of development and commercial integration in the region; the goal is to protect common interests among South American countries from pressures by more economically powerful nations, and also to resist United States' efforts to integrate South American nations as individual entities to NAFTA.[19]

IIRSA was first discussed in a meeting held in Brasília in 2000 following the initiative of President Fernando Henrique Cardoso. Twelve South American presidents participated in this meeting. Cardoso's strategic and diplomatic efforts included: the implementation of a Brazil-Bolivia gasoduct; the import of oil from Argentina and Chile, instead of obtaining it from Arab nations; and the end of the 40 years conflict over border issues between Ecuador and Peru. In Cardoso's own words: "The physical integration of our space has become one of the objectives of the planning proposals taken to be discussed among the presidents of South American countries".[20]

President Luiz Inácio Lula da Silva continued the process of integration of South

American nations with the so-called *diplomacy of generosity* (*consensual hegemony)* which promoted the agreement between Brazil and Bolivia regarding the gasoduct and gas plants built by Petrobras[21] that was later nationalized by Evo Morales. As mentioned earlier, and according to Sean Burges: "The intention was not to seek Brazilian leadership in South America through imposition, but inducing a desire of mutual benefits that could subtly embed Brazilian interests, aspirations and strategies in the region."[22] However, the multiple diplomatic initiatives undertaken by successive Brazilian governments clearly point to a heightened Brazilian leadership on the continent. This includes IIRSA and its multiple developmental projects. As such, IIRSA poses new challenges in terms of the attempted integration of the sub-continent.

Why then is IIRSA a strategic move for Brazil? There are three main reasons: the increasing importance of East Asian countries as trade partners; an increasing interest in the relations between Brazil and other Amazon basin nations seeking access to the Pacific Ocean; and as an important economic and political tool to strength Brazilian leadership in South America. Currently, the building of roads, ports, hydroelectrical power plants, and other infrastructural technological investments, are promoted and financed by major Brazilian companies and investors (mainly transnational agencies based in Brazil: Odebrecht, Camargo Correa, and Andrade Gutierrez). These agencies have been investing heavily in the region, and it is their intention to expand their activities in agribusiness, industries, and biotechnology in the entire Amazon basin. Such explicit intentions replicate the long held dominant role these corporations have had compromising public money and delivering, once again, a high concentration of capital that only benefits a few powerful groups.[23] This is a new form of 'dependency' that perpetuates long held patterns of capital accumulation.

According to Carciofi, "since its launching IIRSA has achieved a series of results while arousing some controversy"[24] because IIRSA has basically ignored the political, social, economic and trade-related dimensions of its road construction projects. Carciofi also argues that South American countries have been involved in continuing with the sub-continental integration through the building of physical infrastructure and the promotion of economic development. Among South American countries there are issues and questions as to the asymmetry of investments to promote infrastructure and development. The remaining question is to what extent will the participating South American countries find a common ground when it comes to protect the environment and the Amazon basin as a complex ecological system?

Strategic Development Hubs

The focus of IIRSA is on infrastructural investments from a regional and transnational perspective: transportation, telecommunications and energy. IIRSA's 2000–2010 Action Plan created ten different axes for development throughout South America, the so called *development hubs*.[25] Each axis encompasses a group of countries with common interests and shared regional characteristics. This initiative aims at contributing to infrastructural integration of the region, that in turn supports the region-wide strategy of *open regionalism*[26], and to strengthen South America's insertion into the world market.

According to the 2005–2010 Consensus Agenda, the total budget amounts to

$10.2 billion to be used mostly to implement transportation. This means that three-fourths of the total amount will be used to build roads.[27] Funding comes from conventional public and private sector institutions and corporations, as well as multilateral financial entities, such as the Inter-American Development Bank (IDB), the Andean Funding Corporation (CAF), and the Financial Fund for the Development of the La Plata Basin (FONPLATA).

Two of the ten development hubs are directly relevant for their impact on the Amazon basin: the *Amazon Hub* and the *Peru–Brazil–Bolivia Hub*. The *Amazon Hub* encompasses an area of 4.5 million square kilometers and a multi-modal transport system interconnecting ports on the Pacific, such as Tumaco in Colombia, Esmeraldas in Ecuador and Paita in Peru, with the Brazilian ports of Manaos, Belen and Macapá (Figure 7.6). To interconnect the Atlantic and Pacific Oceans through waterways, the project also includes the construction of the Pasto–Mocoa highway, connecting the port in Tumaco (Colombia) with the Putumayo and Amazonas rivers; the Taipa–Tarapoto–Yurimaguas highway construction; and the Lima–Tingo Maria–Pucallpa highway, as well as ports such as the Francisco de Orellana Port. As can be seen in the map (Figure 7.6), most of the roads will cut through Amazonean and Andean territories in Peru and Bolivia. Brazil will likely sufffer the least impact

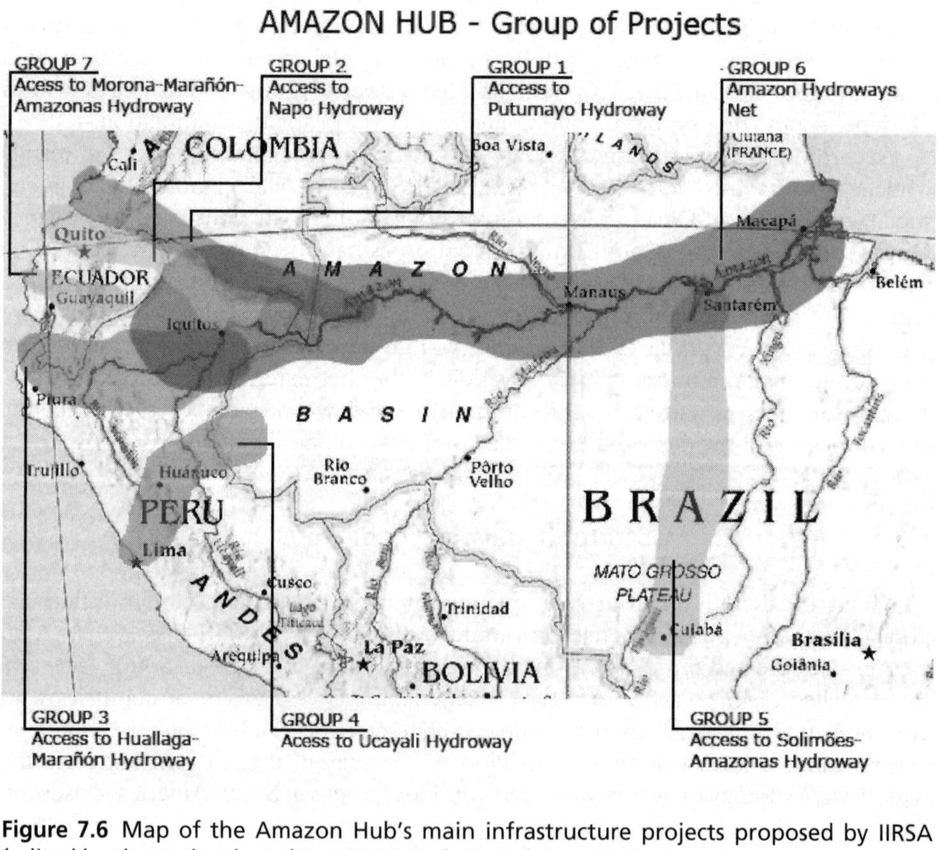

Figure 7.6 Map of the Amazon Hub's main infrastructure projects proposed by IIRSA (edited by the author based on IIRSA's website information).

with the opening of new roads, because the Amazon River is navigable for almost its entire length through Brazil. IIRSA's website reads:

> [The Amazon Hub] development potential shows it is a region with its own very distinct specificities, generating a common identity for all Amazon peoples, which must be the basis for justifying development through the promotion of an Amazon culture to reinforce the appraisal of the natural environment and the environmental services it provides at a global level, and the sustainable use of natural resources through knowledge about the native peoples and science.[28]

The *Peru–Brazil–Bolívia Axis'* main purpose is the paving of the roads Iñapari–Porto Maldonado–Inambari, Inambari–Juliaca, and Inambari–Cuzco, and the building of a bridge over the Acre River to facilitate the crossing of the border. On the Peruvian side, the project consists of widening and paving the existing road between the Inambari River and the Brazilian border and the rehabilitation and improvement of other paved roads through the Andes connecting the ports of San Juan de Marcona, Matarani and Ilo on Peru's southern coast. Also called the "Transoceanic" or "Bioceanic" Highway, this is one of the anchor projects of the Peru–Brazil–Bolivia corridor of IIRSA. It involves the construction and rehabilitation of more than 2,000 kilometers of roads. The Brazilian side of the road is partially constructed (BR-364 and BR-317) and paved from Porto Velho and Rio Branco to the border with Peru at Iñapari, with a total investment of 1.4 billion dollars. Once completed, the highway will facilitate the transportation of Brazilian goods to the Pacific coast (and on to Asian markets) and the transport of Asian products to the Atlantic coast (and on to the US, European and Brazilian markets). Former Peruvian president, Alejandro Toledo, declared that the highway would increase access of Peruvian products to markets in Brazil and that the project would lead to a 1.5 percent annual increase in GDP.[29]

The Amazon Hub may also induce the rise of new centers of economic gravity in the region. As such long-distance hubs extend over a vast territory, they may also foment a large-scale land-use conversion. Will land-use conversions provoke more destruction to the natural environment? Which are the institutional mechanisms and policies for public consultation and the monitoring of IIRSA's projects to avoid the negative impacts of these large-scale, high-risk and debt-heavy mega-projects that result in extensive landscape and livelihood changes in the region?

According to the information on IIRSA's website,[30] the initiative is based on a social and environmental evaluation with a strategic focus and in compliance with national environmental laws. IIRSA's declared principles acknowledge the diversity of the biota and the economic potential of the natural resources in the Amazon basin, including water and minerals.[31]

Strategic Environmental Assessment

The Strategic Environmental Assessment (SEA) is a vital tool to stimulate the potential welfare that might come with infrastructural improvements as it mitigates potential negative impacts for humans and the environment. The concept underlying SEA is relatively new and a follow-up on earlier methodologies to assess infrastructural impacts with costs-benefit analysis (CBA) and environmental impact assessment (EIA). In 2008, BICECA[32] and the Inter-American Development Bank (IDB) discussed the use and misuse of a SEA in a meeting in Bogota, in 2008. After the meeting, IDB produced a SEA Guide in June 2009, which will be applied to future "strategic projects" in the Amazon basin, financed by the Bank.

A recent study on SEA reflects the methodological problems regarding the implementation of SEAs as "the acquisition of information on future developments in the impact area is hampered by poor capabilities to forecast markets and by biases in the results of the public consultation". [33] The total budget for implementing proposals is modest when compared to the overall budget for IIRSA projects. SEA's proposals are reduced to a six months time and US$ 300,000. Pitou van Dijck and Simon den Haak also point out that "earlier experience with road construction in Amazonia shows that unexpected spatially widespread negative environmental and economic developments with an irreversible character have occurred in cases characterized by inadequate [or inexistent] ex-ante assessment".[34] Stakeholders involved in planning and constructing roads and infrastructure have their own expectations regarding future revenues, to a point where one can doubt if with all the budgetary and temporal constraints and impediments SEA can provide guidance to really mitigate or avoid major negative impacts.

So far only a few strategic assessment studies have been carried out related to corridors in the IIRSA agenda. Moreover, only little systematic knowledge and know-how has been developed toward the appropriate methodology for studies of such large-scale corridors.[35]

In Lieu of Conclusions

Among IIRSA's sustainability efforts four main priorities are listed:

> (i) economic sustainability, achieved through the productive processes' efficiency and competitiveness; (ii) social sustainability, granted by the visible impact of economic growth on the quality of life of the population at large; (iii) environmental sustainability, which entails the rational use of natural resources and the conservation of the ecological heritage for future generations; and (iv) political-institutional sustainability, which consists of creating conditions so that society's different public and private agents can and will be willing to contribute to development and integration.[36]

The open question is as to whether "development" and "progress" can be achieved without a high cost paid by the environment and human and social welfare. Natural resources in the Amazon basin are increasingly becoming global treasures. Any threat to these resources may be considered more of a generational challenge than as the purview of any one nation. With the use of SEAs, a more comprehensive and inclusive approach to large infrastructural and regional development projects becomes possible, particularly in the case of border-crossing projects and public-private-partnership constructions. Public and local participation is needed to find a more balanced input of all involved in thinking about the future of the Amazon basin. Currently, the decisions on how and where the investments are made lies with large and private corporations and their economic interests and political intentions. And the corporations' main outlook is profit making in the global market, without much (if any) concerns regarding the environment, biodiversity and social welfare. Diplomacy and a transparent and democratic debate need to be an integral part of any development effort.

Notes

1 Verdum and Faletto, 2008.
2 Cardoso 1979.
3 Borges 2010.
4 Skidmore and Smith 2005.
5 Sean Burges in Borges 2010.
6 Christopher Brown, herein.
7 Cerrados, also known as the Brazilian Savanah, is one of the ecosystems, or biomes, found in Brazil, especially in the States of Bahia, Minas Gerais, Goiás, Mato Grosso and Mato Grosso do Sul. It is characterized by semi-arid climate with drought tolerant vegetation and a significantly rich biodiversity.
8 According to recent studies 63 percent of the Brazilian soy bean is produced in the Cerrado and the expansion of its production is a real threat to this sensitive biome. http://www.brasiloeste.com.br/noticia/1126/expansao-soja and http://www.embrapa.br/imprensa/noticias/2008/outubro/1a-semana/cultura-da-soja-no-cerrado-e-tema-de-reuniao-tecnica/
9 Amayo 2007.
10 Op. cit.: 111.
11 Hayes-Bohanan 2001.
12 SIVAM is the Amazon Surveillance System, a government tool that monitors Legal Amazon through satellite and has among its objectives environmental control, curb drug trafficking, deforestation and log smuggling, and air traffic control.
13 For the satellite images, access the website http://na.unep.net/atlas/webatlas.php?id=29
14 The two peaks of deforestation in Rondônia were 1995 and 2004, when over 4,700 sq kms. were destroyed (1995) and almost 4,000 sq kms were destroyed (2004). See:<http://news.mongabay.com/2010/0131-rondonia.html>
15 My quote, to stress the view of the military government regarding the Amazon basin.
16 The 'Legal Amazon' consists of the Brazilian Amazon basin and includes the states of Amazonas, Pará, Acre, Amapá, Rondônia, Roraima, Tocantins, in the northern region, and most of Maranhão and Mato Grosso in the northeast and center-west Regions respectively.
17 Pitou van Dijck 2008: 101.
18 Interview published in newspaper O Estadão, São Paulo. Published on 11/25/2007. Available at: http://www.estadao.com.br/amazonia/conflito_agrario_terras_ da_discordia .htm. My translation.
19 Amayo 2007.
20 Borges 2010: 5.
21 Petrobrás is a semi-public Brazilian multinational oil, gas and energy corporation headquartered in Rio de Janeiro, Brazil, one of the global leaders.
22 Sean Burges in Borges 2010: 7.
23 Borges 2010 (from the original).
24 Carciofi 2008: 51.
25 Namely, the ten development hubs are: Amazon Hub, Peru–Brazil–Bolivia Hub, Andean Hub, Southern Andean Hub, Capricorn Hub, Central Interoceanic Hub, Guianese Shield Hub, MERCOSUR–Chile Hub, Paraguay–Parana Waterway Hub and Southern Hub.

26 The concept of "open regionalism" was created by CEPAL in the 1990s to promote Latin America's insertion into the world economic globalization process (Corazza, 2006).
27 Pitou van Dijck 2008.
28 http://www.iirsa.org/BancoConocimiento/P/principios_orientadores/principios_orientadores_ENG.asp?CodIdioma=ENG&CodSeccion=117.
29 For more information access the Bank Information Center at http://www.bicusa.org/en/Project.10312.aspx
30 www.iirsa.org
31 Available at http://www.iirsa.org/ejedelAmazonas_ENG.asp?CodIdioma=ENG
32 BICECA stands for Building Informed Civil Engagement for Conservation of Andes-Amazon (http://www.bicusa.org).
33 Dijck and Haak 2008:118.
34 Dijck and Haak 2008:116.
35 http://www.iadb.org/features-and-web-stories/2008-04/english/cooperation-and-provision-of-regional-public-goods-the-iirsa-case-4561.html
36 http://www.iirsa.org/BancoConocimiento/P/principios_orientadores/principios_orientadores_ENG.asp?CodIdioma=ENG&CodSeccion=117

Bibliography

Almeida, Paulo Roberto. 2009. "Lula's Foreign Policy: Regional and Global Strategies". In: Love, Joseph L.; Baer, Werner (eds.) *Brazil under Lula: Economy, Politics and Society under the Worker-President*. United States of America: Palgrave Macmillan, pp. 167–183.

Amayo-Zevallos, Enrique. 2007 "Amazonia, MERCOSUR and the South American regional integration". In: *The Bush Doctrine and Latin America*. Edited by Prevost, Gary and Campos, Carlos Oliva. New York: Palgrave Macmillan, pp. 105–128.

Becker, Bertha. 2007. Reflexões sobre a geopolítica e a logística da soja na Amazônia. In: Becker, Bertha et al. (orgs.). *Dimensões humanas da biosfera-atmosfera na Amazônia*. São Paulo: Edusp.

Borges, Fábio. 2010 *As relações do Brasil com os países amazônicos nos governos de Fernando Henrique Cardoso e Lula (1995–2006): "hegemonia consensual" e seus limites*. Proceedings of VI Congrès du CEISAL. Toulouse: CEISAL.

Carciofi, Ricardo. 2008. Cooperation and Provision of Regional Public Goods: the IIRSA case. In: *Journal of Integration and Trade*, vol. 28. Buenos Aires: Institute for the Integration of Latin America and the Caribbean Inter-American Development Bank — INTAL. Available at http://www.iadb.org/intal/aplicaciones/uploads/publicaciones/i_INTAL_I&T_28_2008_Carciofi.pdf

Cardoso, Fernando Henrique and Faletto, Enzo. 1979. *Dependency and development in Latin America*. Traduzido por Marjory Mattingly Urquidi. Berkeley University of California Press.

Corraza, Gentil. O "regionalismo aberto" da CEPAL e a inserção da América Latina na globalização. In *Ensaios FEE*, Porto Alegre, v. 27, n. 1, pp. 135–152, maio 2006.

Dijck, Pitou van. 2008. "Troublesome Construction: The Rationale and Risks of IIRSA". In: *European Review of Latin American and Caribbean Studies* 85, pp. 101–120.

Dijck, Pitou van and Haak, Simon den. 2006. "Troublesome Construction: IIRSA and Public-Private Partnerships in Road Infrastructure". In: *Cuadernos del CEDLA 20*. Amsterdam: Center for Latin American Research and Documentation (CEDLA).

Hayes-Bohanan, James. 2001. What a difference a century makes: images of the frontier in the Amazon and the American West. In: *Olhares sobre a Amazônia: Looking at the Amazon*. Nenevé and Cooper, eds. Rio de Janeiro: Terra Nova.

Ramos, Leonardo. 2009. "Contra-hegemonia e política externa? A política externa brasileira no governo Lula". In: LASA — Latin American Studies Association XXVIII International Congress. Available in: http://lasa.international.pitt.edu/members/congress-papers/lasa2009/files/RamosLeonardo.pdf.

Skidmore, Thomas and Smith, Peter. 2005. *Modern Latin America*. New York: Oxford University Press.

Verdum, Ricardo (org). 2008. *Financiamento e Megaprojetos: Uma interpretação da dinâmica regional sul-americana*. Brasília: INESC.

Websites

http://www.iirsa.org
http://www.bicusa.org
http://www.iadb.org

8 Debates in the Environmentalist Community: The Soy Moratorium and the Construction of Illegal Soybeans in the Brazilian Amazon

J. Christopher Brown and Matthew Koeppe

Introduction

As the title implies, it used to be legal to plant soybeans in the Brazilian Amazon, as long as national environmental laws governing deforestation were followed. Given fairly low land prices in this relatively remote area of Brazil, and the world market's insatiable appetite for protein, among other reasons we detail below, soybeans have found a comfortable home in what has become one of the world's newest "bread-baskets" (Economist, 2010). The increased environmental and social costs of such rapid expansion, however, led international environmental nongovernmental organizations (NGOs) to take action against it (Howden, 2006). As a result, in 2006, the landscape of Amazon soybean production changed dramatically.

A so-called "soy moratorium" was declared, making it "illegal" to purchase any soybeans grown on deforested land in the Amazon biome after July of 2006 (ABIOVE (Brazilian Association of Vegetable Oil Industries), 2006). This declaration was not made by any court of law, not by any presidential decree. Rather, large agribusiness firms in Brazil, with substantial global corporate and non-governmental environmental organization support, declared the moratorium. With it, they pledged to govern themselves, but in effect, the moratorium made illegal what used to be perfectly legal. There was a dramatic effect on what farmers could plant, where they could plant it, and arguably the moratorium has contributed to lower deforestation rates in the Amazon. The moratorium was not merely another "new law." Its efficacy relied on a geographic information gathering and enforcement system that appears to be an example of neoliberal environmental governance that works, the kind of international market-based conservation opportunity cheered by some as leading to the end of deforestation in the Amazon basin (Nepstad, Soares et al., 2009).

The soy moratorium occurs in the midst of a lively debate in the environmentalist community about how best to conserve forests and other sensitive areas. On one hand, is the emerging group that is calling for more market-led solutions, such as the soy moratorium. On the other hand, are the groups that support a more traditional, state-led approach, centering on the creation and enforcement of protected areas and supplemented with state-led incentive programs that would dissuade people from moving to and clearing areas of forest. In the Amazon, these two contrasting views are well represented by the writings of Daniel Nepstad and Philip Fearnside, respectively. Both have the same ultimate goal — stopping deforestation — but they have come to endorse somewhat different strategies for achieving it.

Fearnside has long been an advocate of an approach that mixes a variety of state-led strategies:

> Removing the motives for deforestation will require a stop to regularizing land claims, as well as cutting the various subsidies through which Brazilian taxpayers still contribute to the profitability of clearing. Controlling deforestation requires establishment of the rule of law throughout Amazonia, including a complete cadaster, or register, of land titles and reinforcement of command-and-control programs. Movement to the frontier must be discouraged by exercising restraint in approving infrastructure projects, especially for highways, and by creating and protecting conservation units. At the same time, economic alternatives must be created by generating employment in source areas and in alternative destinations, by supporting sustainable uses of the forest and alternative supply of national markets for products such as wood. Most fundamental is continued progress in creating mechanisms to reward the environmental services of standing forest as an alternative foundation for the economy in rural Amazonia. (Fearnside, 2008: 16)

In contrast, Nepstad trumpets the power of globalized markets to create opportunities for conservation:

> Brazil has two major opportunities to end the clearing of its Amazon forest and to reduce global greenhouse gas emissions substantially. The first is its formal announcement within United Nations climate treaty negotiations in 2008 of an Amazon deforestation reduction target, which prompted Norway to commit \$1 billion if it sustains progress toward this target. The second is a widespread marketplace transition within the beef and soy industries, the main drivers of deforestation, to exclude Amazon deforesters from their supply chains. (Nepstad, Soares et al., 2009: 1350)

As is evident in these two quotations, where Fearnside sees largely state-led solutions to the problem of Amazonian deforestation, Nepstad sees market-led solutions. Even more broadly, Fearnside looks to internal (domestic) forces, while Nepstad looks to external (international) ones. The key differences between these two approaches are the enforcement mechanisms. Fearnside's approach recognizes the fundamental role of the state in the enforcement of forest policy, executable via fines and related command and control mechanisms. Alternatively, the state can orchestrate incentives in the forms of tax breaks, jobs programs, and the like. Nepstad, in contrast, sees the market as the most practical and effective enforcement mechanism. The logic is as straightforward as state enforcement via protected areas — if farmers cannot sell their products, then they will not undertake efforts to clear forests for farmland. The strategy exemplified by Fearnside's quote represents the status quo for many countries, including Brazil. The strategies endorsed by Nepstad, and identified as "major opportunities to end the clearing of [the] Amazon forest," are newer.

This chapter aims to add a level of richness to our understanding of current efforts to stop deforestation as being either state-led, or market-led. We consider the history of how trading soy came to be "illegal" in the eyes of the market and how market-led strategies have been able to achieve a level of enforcement that the state was never willing or able to accomplish. This history makes it clear that the soy moratorium could not be the market-led success it is if it were not first for the establishment of a very strong state-led effort, beginning over half a century earlier, to map and monitor the resources of the Amazon. Such efforts date back to the 1930 military-backed installation of Getúlio Vargas to the Brazilian presidency and the eventual establishment of the *Estado Novo*, during which Vargas built strong national state institutions and information gathering systems for the very first time. These institutions were essential in building the knowledge base, and the technology to develop, the resources of the Amazon in order to build the new national state. Subsequent military dictatorships during the Cold War further consolidated this trend. By the time democracy and global market forces were in full play at the start of the twenty-first century, hybrid state-market policies finally made it possible to render certain activities "illegal," backed by enforcement powers that no purely state-led or purely market-led strategy could muster alone.

This chapter views the history of the making of illegal soy as going through three main periods, roughly following Brown and Purcell's (2005) analysis of the changing scales of control over decision-making in Amazonian development. The main difference in the writing of this history is that we are focused on how the making of national state territory played an essential role in building the institutions and knowledge required to make new laws and enforce them. This chapter retells this history, albeit briefly and necessarily glossing over much detail, in order to provide some essential context for the soy moratorium as a governance strategy. It is also important to establish the view taken in this chapter that laws, in and of themselves, have territorial properties — some action is allowed here, but not there — and thus they are an important expression of human territoriality (Sack 1986). Thus, our understanding is that new territories can lead to new laws and vice versa, and in fact the establishment of law and territory go hand in hand and cannot easily be separated. In short, the establishment of any territory requires some knowledge of what is to be delimited into a territory along with some justification. Enforcing territorial limits, in turn, requires further information and surveillance, which can aid in an even greater detailed delimitation of territory. Within each period of the making of illegal soy, this chapter focuses on what political economic interests were in control of the Amazon, the institutions and laws that governed land use, and the information required for the enforcement of the law.

President Getúlio Vargas, the Military, and the Making of National State Territory (1930–1964)

Getúlio Vargas came into Brazil's presidency, installed by military force, after a long period of extremely decentralized government known as the Old Republic (1890–1930). During this prior period, regional elites in the most economically vibrant region of Brazil (coffee barons in the southeast) dominated national-level politics, setting policies regarding trade and taxes that benefitted themselves, regardless of the needs of other regions and economies (Weinstein, 1983; Prado Júnior, 1987). This extreme regionalism, also known as "the politics of the states," allowed regional elites elsewhere, such as rubber barons in the Amazon, to control their own governance and the benefits of their own trade, without interference from the national polity (Burns, 1993: 266). Vargas's rise to power is most-often understood as a reaction to this extreme decentralized political system. Once coffee barons in the southeast began to lose their economic power, due to depressed coffee markets, other emerging interests in Brazil saw an opening to gain power under the belief that the regional interests were too selfish (Weinstein, 1983). Backed by industrialists, growing middle and intellectual classes, and idealistic military officers, Vargas rose to power in 1930 with a project to build an authoritarian nationalist state based on the full incorporation of national state territory, people and resources, for the good of Brazil (Wirth, 1970; Bak 1983; Bak, 1985). This national project involved dismissal of regional and municipal legislative bodies, the development of national-level policies and the formation of national political parties. The year 1937 marked the strongest-ever nationalist and authoritarian nature of Vargas's project when he cancelled presidential elections and dismissed congress, declaring his establishment of a new government called the *Estado Novo* (Skidmore and Smith, 2001).

With respect to the Amazon, Vargas specifically called for Brazilian national civilization to spread westward into the region as part of an effort to bring the full Brazilian territory into the political and economic life of the nation. This "March to the West," was built on the view that Amazonia was an unpopulated, vast region with the potential for agricultural and natural resource development (Hecht and Cockburn, 1990). The view resonated strongly with military concerns that neighboring countries and communist insurgencies could wrest control of the region from national control (Vargas, 1938–1941; Silva, 1967). In short, Vargas's project, regarding the Amazon, was to populate it, control it, and harness it as part of building a strong national state. Vargas came in and out of power through this period until committing suicide in 1954. The military, deeply involved in Brazilian politics throughout the period, continued to build on many of the nationalist efforts Vargas originally put in place.

Vargas and the Brazilian military set the stage for a strong national state beginning in the 1930s with key institutions and laws that were consistent with Vargas's project. In 1931, Brazil's forest code was enacted as part of a suite of codes that also regulated scientific expeditions, water use, mines, and hunting and fishing (Dean, 1995). In the code, Vargas's "denial of liberalism" and his effort to build Brazil "under the banner of a modernizing and technocratic nationalism," is clear (Dean, 1995: 261). The code abolishes the absolute right of property, and it set rules regarding tree cutting. Trees must be preserved along waterways to protect watersheds, and when they house rare species. Land owners could not cut more than three-quarters of forested land. In this first iteration, the forest code applied to all of Brazil, for the Amazon had yet to be officially distinguished in Brazilian law (Browder and Godfrey, 1997: 256). This soon began to change as the Brazilian military and other national institutions increasingly took an interest in the region, recognizing its potential importance for national defense and economic development. In 1946, the newly ratified constitution called for a comprehensive long-term plan for the integration and development of the Amazon. In 1953, the Superintendency for the Economic Valorization of the Amazon (SPVEA or *Superintendência do Plano de Valorização Econômica da Amazônia*) was established to develop and carry out such a plan (Hecht and Cockburn, 1990). One of the Superintendency's first acts was to legally define the Amazon region (the Legal Amazon or *Amazônia Legal*) as a form of bio-administrative region. The boundary included the states of Mato Grosso, Goiás, and western portions of Maranhão, thus

Figure 8.1 Map of Brazil showing the humid forest biome, savanna, the Legal Amazon, and the municipality of Vilhena, Rondonia.

humid tropical forest and significant areas of savanna, or cerrado, were part of the same territory (Figure 8.1) (Hecht and Cockburn, 1990; SUDAM, 2011).

One of the key institutions for all these developments during the Vargas years was The Brazilian Geography and Statistics Institute (IBGE or *Instituto Brasileiro de Geografia e Estatística*), which was a 1938, *Estado Novo* renaming of an earlier national statistics institute that Vargas had established soon after taking power in 1930 (Almeida, 2003). Most pertinent to our discussion is the Institute's purpose to amass knowledge and statistics about Brazil's socio-economy and its natural resources for the purposes of providing the information needed for state-led development. It was the Institute's business to help organize the nation's territory, from the national to the local level, and to map the extent of Brazilian territory to aid in government plans to incorporate greater and greater amounts of Brazilian space into the national project. According to Eli Penha (1993), the Institute was an essential part of the government's centralized knowledge gathering activities that helped organize Brazilian territory for nation-building, national policies, and to take the immensity of Brazilian territory and break it up into political-geographic-administrative units. Roberto Almeida (1994) emphasizes the involvement of the IBGE in providing the information needed for the development of major state enterprises under the discourse of national security: from the *Companhia Vale do Rio Doce* (major mining firm) in 1942 to *Petrobrás* (Brazil's National Petroleum Company) in 1952, none of it could have taken place without the mechanisms for territorial control that the IBGE helped put in place. This included systematic knowledge of the physical aspects of the surface and subsurface, infrastructure (existing and planned), defining Brazil's regions, socio-economic understanding of those regions, and mapping.

Military Dictatorship, Development, and the Decade of Destruction (1964–1984)

After Vargas's death in 1954, the military remained deeply involved in Brazilian politics until a coup in 1964 when it took definitive control of the country, opened Brazil's borders to greater international investment, and established a development strategy based on rapid industrialization and greater incorporation of the Amazon into the nation's political and economic life (Stepan, 1977). Some of the main steps to achieve this included the funneling of additional capital to the region; a new forest code was established, and a National Integration Project (PIN or *Projeto de Integração Nacional*) initiated a massive effort to survey the region and its resources and to populate it. These efforts helped further set the territorial limits of investment geared toward Amazonian development, and out of the hands of insurgents who would disrupt the national project. In the process, however, the efforts also began to define the converse of development, which was the requirement that landholders set aside a certain portion of their land for the purpose of conservation.

Based on the Brazilian Geography and Statistics Institute's identification of Brazil's regions, the military established new regional development "superintendencies" to funnel capital specifically to each region. The Amazon's unit, formed in 1966, was called the Superintendency of Amazonian Development (SUDAM or *Superintendência Desenvolvimento da Amazônia*). Soon after taking power, the military established a new forest code. The Forest Code first put into place by the Vargas Administration had gone largely unenforced (Dean, 1995). The new code was enacted in 1965, and its most important innovation was a differential restriction on deforestation, based on whether land is in or outside of a particular area of the country, presumably where humid tropical forests predominate. Recall that the Legal Amazon was first created in 1953, but the 1965 code does not name it specifically. Rather, the key article pertaining to deforestation and conservation, Article 44, states that in the "North region and the northern part of the Center-West region," only up to 50 percent of the tree cover could be removed from a property:

> Art. 44. Na região Norte e na parte Norte da região Centro-Oeste enquanto não for estabelecido o decreto de que trata o artigo 15, a exploração a corte razo só é permissível desde que permaneça com cobertura arbórea, pelo menos 50% da área de cada propriedade. (Lei n° 4771, de 15 de setembro de 1965)

Though ostensibly responding to increasing demands for forest conservation, an underlying purpose of the 1965 Forest Code was to exert oversight and control over large landowners, who were seen by the military as potentially problematic due to their opposition to certain agrarian reforms (Dean, 1995). The Forest Code as written in 1965, though subsequently modified, remains the basic template for forest governance in Brazil up to the present (2011).

With the boundaries of Brazil's Amazon region taking shape administratively and with respect to the new forest code, the military simultaneously initiated an aerial survey of the vast territory of the Amazon, mapping relief, vegetation, geology, soils, and mineral deposits, echoing Vargas's concerns about knowledge of and control over the Amazon as questions of national security. The roots of this effort go to an agency called the National Commission of Space Activities (CNAE or *Comissão Nacional de Atividades Espaciais*), founded in 1963 (INPE, 2011). It was then that Commission personnel travelled to NASA for training in the use of side-looking airborne radar (SLAR) aboard an airplane to create a 1:1.000,000 survey of the nation's natural resources (Oliveira, 1999). This link between NASA and the Commission to survey the resources of such a huge land mass potentially signals that not only was it a project for Brazil's military government to build its version of the national state, but it also served the US's hemispheric interest in ensuring this vast region did not become a communist stronghold during the Cold War. Surveillance of the Amazon on an annual basis also became possible with new satellite remote sensing technologies. In 1972, NASA launched the first of its Landsat program satellites that transmitted imagery back to earth, allowing for the tracking of Amazon deforestation from year-to-year. One year earlier, the military dissolved the old National Commission of Space Activities and formed the present day National Institute of Space Research (INPE or *Instituto Nacional de Pesquisas Espaciais*), a key institution that has over the years produced official deforestation statistics and maps from interpretation of satellite imagery, primarily from NASA satellites that were first launched in the early 1970s.

This Radar Amazonia project (RADAM or *Radar na Amazônia*) became operational in 1970 as part of the larger National Integration Project (PIN) (Foresta, 1991). The project was funded heavily by international capital, and it was followed by a series of development programs funded in part by investments from the World Bank. The project, and the numerous others that followed, allowed for the construction of new major highways, and improvements to older ones, vastly improving connections between the Amazon and the rest of Brazil. They also allowed for colonization of the region by Brazil's poor, which were, until then, largely concentrated in Brazil's south, southeast, and northeast. Tens of thousands of families moved into the Amazon to take part in the private and state-led colonization projects along the new highways beginning in the 1970s. PIN and numerous regional subprojects in later years succeeded in infrastructure improvements and the settling of a whole new generation of Brazilians on the Amazon frontier, helping lead to an Amazon region today that is mostly urban (Browder and Godfrey, 1997). They failed, however, in mitigating the social and environmental costs of this mass movement of people. The so-called "Decade of Destruction" ensued across the Amazon, and by the early 1980s, the Brazilian government and its financiers were accused by Brazilian citizens and international environmental organizations alike, of a number of transgressions including: failing to protect indigenous populations from encroachment by settlers, high deforestation rates, poor delivery of government services to settler populations, and high rates of rural violence associated with contested land tenure on the frontier (Cowell, 1990; Millikan, 1992; Rich, 1994).

In sum, this period was marked by a very intense focus on the Amazon as a major component of a national development strategy, requiring vast amounts of information, vast investments in infrastructure, mass movement of people, and subsequent impacts on people and the environment. Brazil's national, state-led project, though dependent on a huge influx of international and private capital, was about to be challenged by its own people, as the country entered a period of democratization, and by international environmental and human rights groups.

Democracy, Market-led Conservation, and the Soy Moratorium (1985–present)

With democratization spreading in Brazil and around the world in the 1980s and early 1990s, the world entered an era recognized for the so-called weakening of state institutions. Neo-liberalism began to take hold as a dominant economic discourse and practice that shaped state-society relations and had its own impacts on, and opportunities for, environmental conservation (Nepstad, McGrath et al., 2002; Nepstad, Stickler et al., 2006; Brannstrom, 2009). The year 1985 marked an important transition in Brazil, with the presidency of José Sarney, the first civilian president since 1964. He came into office in the middle of the "Decade of Destruction," when world attention on the human suffering and massive deforestation occurring in the Amazon reached its highest level. The push to develop the Amazon in a state-led, military fashion came under attack, often fueled by case studies of abuses and also by the new satellite images from space showing deforestation. (The imagery was interpreted, analyzed, and reported annually by Brazil's own National Space Research Institute, which tracked deforestation across the entire Legal Amazon (Câmara, Valeriano et al., 2006)). International human rights and environmental NGOs exerted tremendous pressure on Brazil and the main financiers of its development (multi-lateral institutions like the World Bank) to change business-as-usual development practice. This pressure is exemplified well in the international NGO campaign against the World Bank and the Brazilian government for their controversial agricultural development projects in the state of Rondonia (Brown, 2001). As a result of this pressure, the World Bank redesigned the approval process for its loans for development in sensitive areas like the Amazon, and governments had to become more sensitive to citizen concerns about protecting human rights and the environment in development.

The Forest Code underwent minor revisions in 1989, but the rule requiring the maintenance of 50 percent of forest cover on landholdings remained in force at the time. In 2001, under increasing pressure from the environmentalist community, the Brazilian government made significant changes to the details of the Code through presidential decree, or what is known as a *medida provisória*. The Forest Code is now referred to informally as the "*Medida Provisória*," or simply "2166," the number of the decree. The important changes to the law decreed in the *Medida Provisória* include: the definition of "areas for permanent preservation" (APP or *Área de Preservação Permanente*), the formal definition of "legal reserve" (which had been included in the 1989 revision), and most importantly, changes to the legal limits of deforestation on private properties in the Legal Amazon. The "APP" is defined as areas that can never be deforested, with very few exceptions, in order to preserve features such as hydrological resources, geological stability, and critical habitats, among others. The "legal reserve" is defined as the percentage of the property that must be maintained under forest cover, not including any APP. The *Medida Provisória* changed the percentages of properties that must be maintained in legal reserve to 80 percent in areas of humid forest, and 35 percent in areas of cerrado, or savanna. On properties that contain both humid forest and savanna vegetation, the percentages of legal reserve are to be assigned proportionally. The legal reserve is just 20 percent in the rest of the country. These new percentages are laid out in the Forest Code's revised article 16 (Medida Provisória° 2166–67, de 24 de agosto de 2001).

These changes in the forest code are extremely important; the code went beyond being specific to the Legal Amazon (one with easily identified boundaries). It set the rules for the percentage of native vegetation required to be maintained in legal reserve according to what type of vegetation existed on landholdings, savanna vs. humid forest *within* the Legal Amazon, a distinction that is not always easy to define, as discussed below. By setting the limit on deforestation to 20 percent of landholdings in humid forest, and up to 65 percent removal of vegetation in areas of savanna, the code explicitly put a greater conservation value on humid forest than savanna, in effect codifying the notion of the savanna as a "sacrifice zone" in Brazil's attempts to develop its vast Amazonian interior (Brannstrom, 2009).

Deforestation for the purposes of agricultural and pasture development continued apace during the 1990s and 2000s, closely tracked by INPE (Câmara, Valeriano et al., 2006). Through this period, the Brazilian government showed itself incapable or unwilling to halt deforestation in command and control fashion, despite the wealth of geographically precise information it had at its disposal to locate offenders of Brazil's forest code, and in spite of some well publicized efforts now and then at the state and federal level to crack down (Fearnside, 2003). Consistent with neo-liberal discourse and practice, there were more and more calls for market-led conservation efforts. This can be seen in arguments for "productive conservation," in which there is a push to increase market demand for products that require the maintenance of forest cover (so-called non-timber forest products), which experienced varying degrees of success and problems (Anderson, 1990; Browder, 1990; Hall, 1997; Brown, 2001).

In spite of these efforts, deforestation continued unabated, linked heavily to increasing market demand for beef, other livestock, and soybeans, produced within the Legal Amazon. By the mid to late-1990s, the state of Mato Grosso had become the center for the cultivation of soybeans on large landholdings and with some of the most modern agricultural technology available in the world (Carvalho, 1999; Fearnside, 2001). Used commonly in vegetable oil, as feed for chickens and hogs, and exported to Europe and China, soy had a major role in building economic and political power in a region that not long before was a backwater. This is best exemplified by the rise of the largest soybean producer in the world, Blairo Maggi, who eventually became the governor of the state of Mato Grosso, and who arguably helped deliver agri-business support for the re-election of President Lula for his second term in 2006 (Economist, 2006).

While researchers in Brazil sounded the warning that market-demand for Amazonian soy was now *causing* deforestation, the international non-governmental organization, Greenpeace, was preparing its case that the same market could be used to stop deforestation, if global consumers only knew the true environmental and human costs of soy production in the Amazon. Greenpeace made the case through two important actions. One was the release of a document entitled, "Eating the Amazon," in April of 2006, which made the detailed case of the manner in which the international consumption of chicken, hogs, and soy was linked to deforestation in the Amazon. The other action was much more direct, in classic Greenpeace style. In May of 2006 at the soybean port of Santarém, Pará on the Amazon River, activists arrived in the ship, "Arctic Sunrise," unfurled spectacular banners, and temporarily blockaded the port. Greenpeace targeted the port for the action because it believed the port, financed and built by the American company, Cargill, was illegal (Greenpeace International, 2006). Some violent altercations ensued between port facility employees and the protestors, all helping the action attract international news interest (Howden, 2006). These actions were successful in getting major soy purchasers and processors, Brazil's vegetable oil industry association, ABIOVE (*Associação Brasileira das Indústrias de Óleos Vegetais*) and Brazil's cereal exporters association, ANEC (*Associação Nacional dos Exportadores de Cereais*), to declare a moratorium on purchases of soy originating from areas deforested in the "Amazon biome" after July 24, 2006 (ABIOVE [Brazilian Association of Vegetable Oil Industries], 2006). All monitoring reports since the moratorium was established show convincingly that only minuscule percentages of lands deforested after 2006 contain soy operations.

This chapter considers the major steps required for the moratorium to have succeeded, steps that a market alone was unlikely to take — steps that required very strong state-led institutions and information building over decades for the purposes of making a strong national entity. While we emphasize state-led phenomena, we are not saying that market-led phenomena were not important. Market-led dynamics were essential, *along with* the state-led ones. With a relatively weakened state during this neo-liberal period, space was opened for the market and for international NGOs to pull information and economic levers to obtain compliance with moratorium rules. At the same time, however, the moratorium succeeded because of decades of successive governments, beginning with that of President Vargas, which built the sense that the Amazon biome was important to protect; a vast region like this came to be identified as a territory in need of planning, investment, development, and of monitoring and surveillance. In addition, the institutions and technology required to enforce the moratorium, as will be shown, have their roots in state-led nation building, much less in market building.

For the moratorium to work, it had to have some type of institutional organization. Operating first outside of any government agency, and with no official government involvement whatsoever, the moratorium formed a Soybean Working Group, comprised of industry/business interests and several international and local NGOs, including the World Wildlife Fund, The Nature Conservancy, The Institute for Amazonian Research, Conservation International, Greenpeace, and the Amazonia Defense Front, based in Santarém, Pará, the site of the disputed Cargill port facility (ABIOVE [Brazilian Association of Vegetable Oil Industries], 2006a; ABIOVE [Brazilian Association of Vegetable Oil Industries], 2006b). The Working Group divided into three subgroups: a farmer education campaign group, a public relations group designed to build greater international business and NGO support and to begin developing federal government support, and a group to develop a monitoring strategy so that the moratorium could be enforced. As evidence of the lack of initial government involvement in the moratorium, the federal government did not become an official participant in the moratorium until June 2008 (ABIOVE [Brazilian Association of Vegetable Oil Industries], 2008).

Like any law, the moratorium had to define its territorial focus — what was allowed, not allowed, and where. The "Amazon biome" was the region chosen as the geographic target of the moratorium, and for this, the moratorium relied on the distinction between the humid tropical forest biome of more equatorial Amazonia and the drier tropical forest and savanna biomes located to the south (Figure 8.1). Recall that this distinction was first made and mapped with the military's Radar Amazon (RADAM) project of the 1970s, and that President Vargas's Brazilian Geographic and Statistics Institute of the 1930s was charged with keeping this information for planning purposes. Also, the distinction between these two major biomes within the Legal Amazon was first made in the Forest Code in 2001. A much greater percentage of forest could be removed on lands identified as a type of savanna than humid forest. The moratorium, in effect, could rely on these definitions and the fact that a line dividing the two biomes had already been made decades earlier. With the territorial focus defined, the Working Group had to determine what areas were deforested within the Amazon Biome each year and then verifying each year whether soy had been planted there. For this, the moratorium relied on the National Institute for Space Research's annual mapping of deforested areas in the Amazon, and the hiring of a third-party research firm to fly over the identified deforested areas to look for soybeans.[1] In July 2010 the soy moratorium was renewed for the fourth time, and in every monitoring report produced since the beginning, only a negligible percent of area deforested in the Amazon can be attributed to the planting of soybeans. In 2009, Carlos Minc, then-Minister of the Environment, showed his enthusiasm for the moratorium in an announcement that soybean agriculture no longer factored into deforestation (GLOBALSAT and ABIOVE [Brazilian Association of Vegetable Oil Industries], 2008 and 2009; Leão, 2009; Lovatelli, 2009).

Conclusions and a Critique of the "Success" of the Soy Moratorium

The preceding paragraphs placed the soy moratorium into its historical context, providing a much richer understanding of how soybeans came to be "illegal" in the Amazon. The moratorium, as an attempt to stop deforestation around the production of a valuable domestic and export commodity, came at a time when a democratized Brazilian state was bypassed by powerful international agribusiness and environmental non-governmental organizations to force farmers to change their land-use practices. In order for the moratorium to be a success, it required the institutions of a strong national state dating back to the 1930s, and the information these institutions produced and held for government planning. Thus, neither Fearnside's call for strong state-led command and control to stop deforestation, nor Nepstad's faith in global markets to work in favor of conservation are based on a historical, political-economic and institutional understanding of Brazil.

Now considering in more detail the feat that the moratorium achieved, Brazil had never systematically enforced its forest code since it was first formally written in the 1930s. On the one hand, how could Brazil realistically enforce it? The Amazon basin itself is of continental size, and even though Brazil had the information it needed to identify offenders, government officials were loath to make the political and economic investment needed to enforce it. In fact, from the Vargas Administration through the military years, forest code enforcement seemed to be merely a lever that federal government officials could pull at will in order to attack any potential regional landholding opponents of the strong national state project. This is the point at which the neoliberal market has made a major contribution to stopping deforestation. It is a force with a reach so wide that it made it possible to enforce an even more stringent rule than the forest code itself. The moratorium, in effect, made it illegal to do something that was perfectly legal under Brazilian law. It was legal to plant soybeans on deforested land, as long as farmers remained in compliance with the forest code and its restrictions on total area cleared. It is very unlikely that the federal government ever would have passed such a restriction on land use. The enforcement appears to have worked, because farmers received visits by the soy purchasers themselves, signing documents they would not try to sell soybeans from recently deforested land, and that if caught, they would lose their only outlet to the market. They would also lose one of the most important sources of credit for seeds and fertilizers, as the soy purchasers themselves allowed farmers to borrow on the value of the following year's production (Brown, personal communication with government official). Both the purchasers and the farmers have calculated up to now that it would be too costly not to follow the moratorium directives. But again, it is difficult to imagine that a market would have created the kinds of institutions and information the moratorium required. Instead, the moratorium appears to have taken advantage of institutions and information that were first created to build the Brazilian national state, an entirely different purpose.

With the richer historical understanding of the success of the moratorium, we now turn to a critique of the so-called "success" of the moratorium. Despite the positive reports that soy is no longer a factor in deforestation, we do not see the moratorium as a sustainable long-term solution to deforestation in the Amazon. Much of this chapter'scritique is based on the fact that the moratorium is very commodity- and national biome-centric. This chapter concludes by considering each of these limitations in turn. By focusing in on a single commodity, soy, the moratorium makes deforestation out to be the product of a very simple agricultural activity, masking the complexity of known cropping practices in Amazonian commercial agriculture. It is known that soybean is not always the first crop planted after deforesting a plot of land. Upland rice is often planted as a first crop as a step in the process of preparing the soil for soybean and other crops (Brown, Koeppe et al., 2005). One explanation for planting rice first is that rice is harvested higher off the ground, saving harvester equipment from damage by any remaining pieces of wood and stumps in the field. The success of the moratorium thus could be simply because farmers normally do not plant soy anyway on recently deforested land.

The moratorium thus does not stop deforestation for other crops or activities, and pasture expansion is one of these. The market for Amazonian beef is rising, as is demand for cattle hides used in the shoe industry, hence the need for more pastureland. In recognition of this, it is not surprising that Greenpeace and others have spearheaded an effort to develop a moratorium on the purchase of cattle raised on recently deforested land (Greenpeace International, 2009). It is beyond the scope of this chapter to provide an evaluation of that effort, but we mention it to make the point that somewhere along the line, such commodity-centric moratoria, one after the other, run the risk of becoming old and stale, leading to apathy among consumers and a sense among agricultural interests that there could eventually be a return to business-as-usual without fear of any consumer or market backlash. The need to govern the agricultural system as a whole, involving the complex relationships between land clearing, livestock and crop production, is required for long-term sustainability of land-use rules and laws.

Furthermore, the moratorium can be critiqued for its national, biome-centric approach. It makes concerns about global commodity production and environmental degradation out to be something that is particularly Brazilian and found only in Brazil's humid tropical biome. We wonder what effect the moratorium has had on the movement of today's globalized agricultural operations. It is possible that the moratorium has simply "squeezed" operations to nearby Bolivia or south to Paraguay (Kaimowitz and Smith, 2001). Natural vegetation is being cleared, potentially for soy production, in these countries that do not have the kinds of institutions and information required for a similar moratorium to be put in place. Moreover, by focusing on humid tropical forests, it further constructs dry forests and the vast savannas of South America to be sacrifice zones in the battle to achieve some balance between development and environmental conservation (Brannstrom, 2009). We recognize that in any law or rule, a territory must be identified where the rule is in force. The Amazon biome (humid tropical forest) was chosen for the moratorium, leaving the savanna outside the zone of protection. While maps show clear divisions between the two biomes, these are ultimately artificial lines. This is because at the borders of major natural vegetation forms like this are areas of transition, ecotones, or as called in Brazilian vegetation studies, areas of "ecological tension." It is unclear in the monitoring methodology published thus far how farmers who find themselves along this area of transition are treated in the moratorium. The map used to separate the Amazon biome (humid forest) from the savanna is drawn at a scale of 1:1.5 million, which leaves a very large area at the margin where confusion may prevail about whether one is in or outside the enforcement zone.

All in all, a biome-centric approach is extremely difficult to enforce without some confusion. This is what we know from our fieldwork in the municipality of Vilhena, Rondonia, which is an area near the border between savanna and moist tropical forest (Figure 8.1). Recall that in the Legal Amazon, the Forest Code, currently the *Medida Provisória* 2166, distinguishes only between "forest" and "savanna". Under the *Medida Provisória*, only 20 percent of forested properties may be cleared, while 70 percent of savanna areas may be cleared. In order to clear natural vegetation on one's property, one must obtain a license, based on a survey that declares what types of vegetation are present, and therefore how much land may be cleared, issued by the proper authority. In theory, on properties with both forest and savanna, licenses would be granted for clearing proportional to each vegetation type present. The subjectivity of the process becomes evident in municipalities like Vilhena at the ecotone between savanna and humid tropical forest. In these areas, clear distinctions between what is forest and what is savanna are not evident, and perhaps not possible. Within the ecotone, some areas are clearly savanna, some are clearly forest, and between these exists an entire spectrum of mixed or transitional types. Given the binary classification system under the Forest Code, choosing to classify a transitional area as either forest or savanna can depend largely on one's political economic interests and motivations.[2]

In Vilhena, the authority to grant licenses originally lay with the *Instituto Brasileiro do Meio Ambiente* (IBAMA), the federal environmental agency. Officials at the regional IBAMA office in Vilhena tended to classify mixed areas with trees as forest, thereby maximizing the amount of land under conservation. In the early 2000s, the authority to grant licenses passed to the state-level environmental agency, the *Secretaria de Desenvolvimento Ambiental* (SEDAM), or the Secretariat of Environmental Development. Because the Rondonia state government viewed agricultural development in the state as a big economic opportunity, SEDAM officials tended to classify transitional areas, even ones with tall trees, as savanna, much to the delight of farmers (Koeppe, 2005: 110–112). Farmers and landowners also informally tend to identify the vegetation in the region as mostly or all savanna. They do, however, distinguish among a variety of types of savanna, including *campo aberto*, which is essentially grassland; standard savanna; and *cerradão*, or "big savanna." *Cerradão* comprises most transitional types and is the most contentious area for classification. *Cerradão* is characterized by relatively tall but skinny trees and still significant underbrush (Koeppe, 2005: 112). In the end, such confusion may open the door all along the humid forest/savanna transition zone for farmers, soy purchasers, and government officials to resolve disputes informally through some form of clientelism.

To conclude, the soy moratorium in the Brazilian Amazon, and other similar efforts, are often praised as a sign that globalized markets can have positive conservation outcomes. We should be very careful not to reduce the apparent success of these initiatives solely to the market, ignoring the geo-political and historical, state-led dynamics that may have had a major role, as in Brazil, in making them possible in the first place. If we are not careful, we run the risk of attempting to apply moratorium-like efforts in areas that are entirely unprepared or unsuitable for anything like the soy moratorium to function. One thing we must do is shy away from extreme positions calling only for state-led or market-led intervention. The history of the soy moratorium shows that it is neither; one could not have achieved results without the other.

Notes

1 There is clear collaboration between the National Institute for Space Research and the Brazilian Vegetable Oil Industry Association to develop more sophisticated methods for detecting soybean cropping in deforested areas, as evidenced in a recent journal publication (Rudorff et al., 2011). This is an interesting development of the Institute's geographic focus of land change research *vis-à-vis* the Legal Amazon. Normally, the focus has always been on tracking deforestation within the humid forest biome. The soy moratorium appears to have created an interest and need to track land-use practices after deforestation as well. Such a shift has begun to incentivize greater contact and collaboration between the Institute and Brazil's federal agriculture research agency, EMBRAPA (Brown, personal communication with EMBRAPA official).

2 The *Medida Provisória* has been heavily criticized by environmentalists and farmers alike. Environmentalists lament that the law is often either unenforced or ignored. Given the scale of the Amazon and the expansion activities taking place there, the Brazilian federal government lacks the resources to enforce the law effectively. Federal versus state jurisdiction over enforcement has also been in dispute. In some cases where the law is enforced, large land owners may simply deforest however much they like, pay the requisite fine, and view it as a mere business expense (Koeppe, 2005). Many farmers, on the other hand, find the law too restrictive and suggest that conservation should not be carried out on a property-level basis.

References

ABIOVE (Brazilian Association of Vegetable Oil Industries). 2006a. "First meeting on Soy Moratorium." from http://www.abiove.com.br/english/sustent/ms_1reuniao_16nov06 _us.pdf

ABIOVE (Brazilian Association of Vegetable Oil Industries). 2006b. "Soybean Working Group Meeting." from http://www.abiove.com.br/english/sustent/ms_minuta_ata_ 19dez06_us.pdf

ABIOVE (Brazilian Association of Vegetable Oil Industries). 2008. "SOY MORATORIUM: Advances and Next Steps." from http://www.abiove.com.br/english/sustent/ms_relatorio_ministro_7out08_us.pdf

Almeida, Roberto S. de. 1994. "O IBGE e a evolução da idéia de "desenvolvimento" no Brasil." *Revista geográfica* 120 (jul./dic.): 5–26.

Almeida, Roberto S. de. 2003. "A geografia do IBGE: um esboço histórico." *Revista do Instituto Histórico e Geográfico Brasileiro* 164 (418): 79–100.

Anderson, Anthony, Ed. 1990. *Alternatives to Deforestation: Steps Toward Sustainable Use of the Amazon Rain Forest.* New York: Columbia University Press.

Bak, Joan L. 1983. "Cartels, Cooperatives, and Corporatism: Getúlio Vargas in Rio Grande do Sul on the Even of Brazil's 1930 Revolution." *Hispanic American Historical Review* 63 (2): 255–275.

Bak, Joan L. 1985. "Political Centralization and the Building of the Interventionist State in Brazil: Corporatism, Regionalism and Interest Group Politics in Rio Grande do Sul, 1930–1937." *Luso-Brazilian Review* 22 (1): 9–25.

Brannstrom, Christian (2009). "South America's Neoliberal Agricultural Frontiers: Places of Environmental Sacrifice or Conservation Opportunity?" *Ambio* 38 (3): 141–149.

Browder, John O. 1990. "The Limits of Extractivism." *BioScience* 42 (3): 174–182.

Browder, John O. and Brian J. Godfrey (1997). *Rainforest Cities: Urbanization, Development, and Globalization of the Brazilian Amazon.* New York: Columbia University Press.

Brown, J. Christopher. 2001. "Responding to deforestation: Productive conservation, the World Bank, and beekeeping in Rondonia, Brazil." *The Professional Geographer* 53 (1): 106–118.

Brown, J. Christopher, Matthew Koeppe et al. 2005. "Soybean production and conversion of tropical forest in the Brazilian Amazon: The case of Vilhena, Rondonia." *Ambio* 34 (6): 462–469.

Brown, J. Christopher and Mark Purcell. 2005. "There's nothing inherent about scale: political ecology, the local trap, and the politics of development in the Brazilian Amazon." *Geoforum* 36 (5): 607–624.

Burns, E. Bradford. 1993. *A History of Brazil.* New York: Columbia University Press.

Câmara, Gilberto, Dalton. de M. Valeriano et al. 2006. *Metodologia para o Cálculo da Taxa Anual de Desmatamento na Amazônia Legal.* São José dos Campos: INPE.

Carvalho, Renata. 1999. "A Amazônia rumo ao 'ciclo da soja'. Amazônia Papers No. 2.", from http://www.amazonia.org.br.

Cowell, Adrian. 1990. *The Decade of Destruction: the Crusade to Save the Amazon Rain Forest.* New York: Henry Holt.

Dean, Warren. 1995. *With Broadax and Firebrand: The Destruction of the Brazilian Atlantic Forest.* Berkeley: University of California Press.

The Economist. 2006. Lula the political prizefighter. October 26.

The Economist. 2010. The miracle of the cerrado; Brazilian agriculture. 396:58.

Fearnside, Philip M. 2001. "Soybean cultivation as a threat to the environment in Brazil." *Environmental Conservation* 28 (1): 23–38.

Fearnside, Philip M. 2003. "Deforestation control in Mato Grosso: A new model for slowing the loss of Brazil's Amazon Forest." *Ambio* 32 (5): 343–345.

Fearnside, Philip M. 2008. "The Roles and Movements of Actors in the Deforestation of Brazilian Amazonia." *Ecology and Society* 13 (1).

Foresta, Ronald A. 1991. *Amazon Conservation in the Age of Development: The Limits of Providence.* Gainesville, University of Florida Press, Center for Latin American Studies.

GLOBALSAT and ABIOVE (Brazilian Association of Vegetable Oil Industries). 2008. "REPORT: MAPPING AND MONITORING OF THE SOY MORATORIUM." from http://www.abiove.com.br/english/sustent/ms_relatorio_mon_abr08_us.pdf.

GLOBALSAT and ABIOVE (Brazilian Association of Vegetable Oil Industries). 2009.

"Report: 2nd Year of the Amazon Biome Soy Moratorium Mapping and Monitoring." From http://www.abiove.com.br/english/sustent/relatorio08/moratoria_carderno_2monitora_0409_us.pdf.

Greenpeace International. 2006. "Eating up the Amazon." from http://www.greenpeace.org/raw/content/international/press/reports/eating-up-the-amazon.pdf.

Greenpeace International. 2009. Slaughtering the Amazon. Amsterdam.

Hall, Anthony. 1997. *Sustaining Amazonia: Grassroots Action for Productive Conservation.* Manchester, UK: Manchester University Press.

Hecht, Susanna B. and Alexander Cockburn. 1990. *Fate of the Forest: Developers, Destroyers, and Defenders of the Amazon.* New York: HarperCollins.

Howden, Daniel. 2006, 7/26/06. "Eating the Amazon: The fight to curb corporate destruction." *The Independent*, from http://news.independent.co.uk/environment/ article1181617.ece.

INPE. 2011. "História." Retrieved March 27, from http://www.inpe.br/institucional/historia.php.

Kaimowitz, David and Joyotee Smith. 2001. Soybean technology and the loss of natural vegetation in Brazil and Bolivia. *Agricultural Technologies and Tropical Deforestation.* A. Angelsen and D. Kaimowitz. Wallingford, Oxon, UK; New York, NY: CABI Pub. in association with Center for International Forestry Research.

Koeppe, Matthew T. 2005. *Environment, Urbanization, and the Rise of Mechanized Agriculture in Southeastern Rondonia, Brazil.* Ph.D., University of Kansas.

Leão, Lucia. 2009. "Soy moratorium reduces plantings on new deforestations to less than 1%." from http://www.abiove.com.br/english/sustent/imm_nota_ministerio_abr09_us.pdf.

Lovatelli, Carlo. 2009. "Soy Moratorium in the Brazilian Amazon Biome: A successful multi-stakeholder initiative in the fight against deforestation." from http://www.abiove.com.br/english/palestras/cop15_palestra_15dez09_us.pdf.

Millikan, Brent H. 1992. "Tropical Deforestation, Land Degradation, and Society — Lessons From Rondonia, Brazil." *Latin American Perspectives* 19 (1): 45–72.

Nepstad, Daniel, David McGrath et al. 2002. "Environment — Frontier governance in Amazonia." *Science* 295 (5555): 629.

Nepstad, Daniel, Britaldo S. Soares et al. 2009. "The End of Deforestation in the Brazilian Amazon." *Science* 326 (5958): 1350–1351.

Nepstad, Daniel, Claudia M. Stickler et al. 2006. "Globalization of the Amazon soy and beef industries: Opportunities for conservation." *Conservation Biology* 20 (6): 1595–1603.

Oliveira, Ailton. A. B. de. 1999. "Projeto RADAM Brasil: História." Retrieved March 27, 2011, from http://www.projeto.radam.nom.br/ingles/history.html.

Penha, Eli. 1993. "O IBGE: a organização do espaço geográfico brasileiro." *Revista Brasileira de Geografia* 55 (1–4): 147–155.

Prado Júnior, Caio. 1987. *História Econômica do Brasil.* São Paulo: Editora Brasiliense.

Rich, Bruce. 1994. *Mortgaging the Earth: The World Bank, Environmental Impoverishment, and the Crisis of Development.* Boston, Beacon Press.

Rudorff, Bernardo F. T., Marcos Adami et al. 2011. "The Soy Moratorium in the Amazon Biome Monitored by Remote Sensing Images." *Remote Sensing* 3: 185–202.

Sack, Robert. 1986. *Human Territoriality: Its Theory and History.* London: Cambridge University Press.

Silva, Golbery do Couto e. 1967. *Geopolítica do Brasil.* Rio de Janeiro: J. Olympio.

Skidmore, Thomas E. and Peter H. Smith. 2001. *Modern Latin America.* New York: Oxford University Press.

Stepan, Alfred, ed. 1977. *Authoritarian Brazil: Origins, Policies, and Future.* New Haven and London: Yale University Press.

SUDAM. 2011. "Legislação sobre a criação da Amazônia Legal." Retrieved March 27, 2011, from http://www.sudam.gov.br/index.php option=com_content&view=category&layout=blog&id=61&Itemid=36.

Vargas, Getúlio. 1938–1941. *A Nova Política do Brasil / Getúlio Vargas. Da Aliança Liberal às Realizações do 1o Ano de Governo, 1930–1931*. Rio de Janeiro: J. Olympio.

Weinstein, Barbara. 1983. *The Amazon Rubber Boom, 1850–1920*. Stanford: Stanford University Press.

Wirth, John D. 1970. *The Politics of Brazilian Development 1930–1954*. Stanford: Stanford University Press.

PART III

Genocide and Plunder

9 Genocide by Plunder: Indigenous peoples of Peru's Amazonia Confront Neoliberalism

Stefano Varese

Note of Context

In mid-June of 2009, a year-long peaceful resistance movement by the *Awajún* and *Wampis* indigenous peoples of North Western Peru's Amazonia was brought to a tragic end. The invasion of their territories by corporations and the state, and the premeditated and treacherous special forces' attack put an end to the efforts by the Indian and local people who were blockading the Bagua highway. The surprise police attack on the demonstrators took place in the early morning, just a few hours before an official government mission was supposed to meet to peacefully resolve the conflict with the Indian protesters. An undetermined number of demonstrators were killed, wounded, and disappeared. Local Catholic Church officials denounced the subsequent establishment of secret detention centers and the reports that army helicopters had thrown the bodies of killed demonstrators into rivers. In response to the massacre, a group of *Awajún* and *Wampis* who had occupied an oil field operating within Indian territory, killed more than twenty policemen who were guarding the property belonging to a foreign oil company. These horrendous tragedies and loss of human life could have been avoided had the government of President Alan García honored its commitment to meet with the indigenous protesters and negotiate a peaceful resolution to their demands. Needless to say, the mainstream media (re)presented the events in such a distorted way that the *Awajún* and *Wampis'* protesters turned out to be the proverbial barbaric savages that oppose civilization, modernity, and the good intentions of national and transnational corporations and the government to bring wealth and happiness to the poor ignorant Indians.

The Neoliberal Pillage

The great English geographer David Harvey has stated that in this late period of neoliberal and global capitalism, the process of accumulation of capital takes on a new form, which he calls "*accumulation by dispossession*", that is to say *accumulation by plunder or accumulation by pillage.* Accumulation by plunder is the engine of neo-*liberalism that aspires to commodify all the elements of nature, of the world, and of the universe such that all resources, lands, waters, air, animals, plants, minerals, landscapes, and above all, people, can and should be commodified. They have a price, and capital could take them and dispose of them for its own benefit and profit.

Even the disruption — or collapse — of Wall Street that occurred in 2008, can be understood as the intentional result of "accumulation by dispossession." The financial hooks and cheap, and ultimately un-payable, mortgages were offered to the domestic proletariat of the United States with the full knowledge of bankers and moneylenders that the buyers were going to default. The premeditated and astronomical increase in interest rates were carefully hidden in the initial agreements so that the financial institution, the lender, could repossess the house and the first mortgage payments from the victims of this pillage. The financial and "legal" traps of these armed robberies of neoliberalism, in the case of the United States, are made invisible by established technologies of communication that make up part of the financial and well-established banking conglomerates, ever more monopolistic and in connection with the U.S. federal government. The general public struggles to comprehend what exactly has occurred and why the financial and credit systems suddenly stopped functioning smoothly.

Occasionally, nevertheless, even the self-censorship of the media mistakenly disseminate some of the horrible facts produced by this economic and social system. A New York Times article dated October 11, 2009, ran a story about the unquantifiable amount of bodies of deceased family members that are abandoned at morgues and in local hospitals because people do not have the money to pay for a burial or cremation. I suspect that not even in the worst moments of the Middle Ages in Europe did it get to these extremes of misery and desolation. The death of a loved one is turned into merchandise for the private funeral agency and for the cemetery (which is almost always subcontracted by the city to private agencies). Here the *accumulation by plunder* is stripped away of its mask of hypocritical dignity and it is announced on commercial television spots, perfectly filmed, so that it is essential that all individuals of the third age begin to pay comfortable monthly fees for their own burial.

The previous few sentences comment on these manifestations of neoliberalism in the U.S. because they are essential components of the socio-economic model and ideological paradigm that are daily imposed on us by the media as our best and only individual and collective salvation. The Peruvian elite and its government of turn – in this case, the rightist Alianza Popular Revolucionaria Americana(APRA) party of Alan García – are exercising the servile adoption of this model and the authoritarian policy implementation that it requires. Considering the enthusiasm with which the government of Alan García endorsed and imposed on the Peruvian people the signing of the Free Trade Agreement with the U.S. when, more than a decade after a similar agreement signed between the U.S., Canada and Mexico, is showing the destructive economic and environmental effects that it has on the indigenous and *mestizo* small farmers, on the working class, and on the weak middle class of Mexico. Millions of Mexican people and millions of hectares of lands are paying a terrible price for the shameful enrichment of a few thousand national and transnational oligarchs. The question for us, the citizens of an increasing globalized community of workers-producers-consumers who are either territorially dispossessed (people without a land base) or in partial control of lands and resources –as in the case of Amazonia's indigenous peoples, is this: What is the response to this ideological barrage of lies and how to counteract and oppose politically this destructive mode of social and cultural development?

The Masked Genocide

Looking back, one of the more obvious successes of the Euro-American colonialism imposed on indigenous peoples across the continent from the sixteenth century onward was the construction of a widespread ideological discourse that masks before the eyes of the world, the real history of the holocaust of original peoples of the Americas. During the first hundred years of the Spanish military occupation, 95 percent of the indigenous peoples of the Americas disappeared. Conventional and bacteriological warfare coupled with environmental or ecological war, resulted in the extermination of millions of indigenous peoples, who for millennia had flourished on the continent and built sophisticated and complex forms of civilization. Recent studies of historical demography illuminate this horrific period of European expansion and the true cost that indigenous humanity had to pay for the disproportionate growth of Spain, Europe, and thereafter North America. No history texts and very few conventional schools of social thought have called these massive massacres, these killings of men, women, children, elderly, by their true name: genocide. If more than 90 million deaths caused by World War II deserve the tragic name of genocide and the six or more million Jews and other 'undesirables' (Eastern Europeans, homosexuals, and gypsies) annihilated by Nazi-fascists are remembered as innocent victims of the Holocaust, nothing comparable has been remembered, written, nor commemorated for the more than 300 million indigenous people who died to enrich the European monarchies and colonial, and neo-colonial oligarchies. The persistent silence of Western and national intellectuals about the painful and bloody beginning of the life of the West in the Americas, marks the successive systematic alienation of the national societies before their own indigenous peoples who were able to survive and coexist with the new landlords.

By the mid-1960s, French Anthropology introduced the term 'ethnocide' to define the policies and practices of aggression and violence conducted by governments and the private sector towards indigenous peoples. The *Bari* of Colombia, killed like animals by mestizo colonists, became the emblem of this new typology of neo-colonialism. The *Bari* followed the revelations about the *Guayaki* ethnocide of Paraguay, the *Nambikwara* in Brazil and hundreds of other indigenous groups whose existence was not known outside the circles of Anthropology. The absurdity of all these initial movements of "Urgent Anthropology" is that they were not as interested in the life of the natives as they were with the registration and documentation of their cultures and languages which were in danger of extinction. It was not enough that entire indigenous groups were decimated or exterminated for national communities, governments, and academics to call such atrocities by their true name; we were passively witnessing the genocide of entire indigenous populations without whispering a word of compassion. Any dictionary defines genocide as "the deliberate and systematic destruction of an ethno-cultural, political, or religious group." The methods of destruction of a human group, and how numerous the group must be, do not constitute obstacles in classifying the fact as genocide. The hundreds of *Nambikwara*, or the thousands of *Witoto* of the Amazon exterminated by the Peruvian rubber workers at the end of the nineteenth century are victims of genocide just as much as the 800,000 Tutsi of Rwanda or the millions of Armenian victims of the Turco-Ottoman state.

In 1971, a group of Latin American anthropologists gathered on the Caribbean Island of Barbados to debate the situation of the indigenous peoples of the lowlands of Central and South America. The Declaration of Barbados denounced the systematic violence, plunder, and armed aggressions against the indigenous peoples who were subjugated. This caused some reactions between the missionaries and the anthropologists in particular, although there was not one single response from the Latin American governments that continued to ignore the situation of ethnogenocide that has occurred on the hinterlands of their countries. Only one government responded promptly challenging these accusations. The military dictatorship of Uruguay immediately seized all copies of the book of Barbados I — published in Spanish precisely by an Uruguayan editorial — and in the best Nazi-fascist tradition, sent them to be burned. An English edition was saved because it was published in Switzerland. How do we interpret this zeal of military censorship in Uruguay? Without even realizing it, the military dictatorships of South America, culminating with the 1973 coup état in Chile by General Augusto Pinochet assumed the role of official sentinels of Western history. They assumed the role of guardians of civilized Western Christianity, whose higher expressions of civilization have been that they knew how to assimilate and integrate (or disappear) the indigenous peoples of the continent that lived in barbarity and in darkness. Nobody in Latin American countries, especially in Perú, wants to admit that there is a sustained underground current of systematic political genocide of indigenous villages. In an article that I published in *Amaru* (July–September Vol. 3) in 1967 I denounced the government of President Fernando Belaúnde Terry for ordering airplanes from FAP (Peruvian Air Force) to bombard with *Napalm* fire-bombs the *Matsés* community in the Alto Yaquerana region. To carry forth this civilizing action, the 'democratic' government of President Belaunde asked the International Petroleum Company (a branch of the Standard Oil Company of New Jersey) for assistance so that, American engineers and technicians could train the Peruvian military on how to construct napalm fire-bombs. These popular Napalm fire-bombs were the same ones that the United States massively used, a bit later, during the Vietnam War. The bombings were executed by the Fuerza Aérea del Perú (FAP) with logistical help from U.S. helicopters specially procured from Panama, the center of military operations for the U.S. Southern Command. We must remember that the early opening of the School of the Americas was organized and implemented during these same years with the sole purpose of training 'counter-insurgency' technicians, most of them from Latin America.

What to make of the event? With what hypocritical metaphor will we hide this barbarism? What else is needed from a government, a national state and its entrepreneurs to respond to the charge of genocide? How many more atrocities need to be committed by Latin American governments to make these political practices subject to the international war crimes tribunal and become part of the international legal lexicon?

The sum of these scandalous politics of ethno/genocide came to have such large repercussions in the European public opinion, that in 1981, the Bertrand Russell Foundation cited the Fourth International Tribunal Bertrand Russell on the Human Rights of the Indigenous Nations. In Rotterdam, the tribunal and its international jury met, heard, and revised a sample of more than a thousand cases of deliberate forms and systematic acts of ethno/genocide committed against the indigenous nations of the Americas by the governments and private companies in question. The moral condemnation of the Russell Tribunal may not have caused any worry among the governments, but it did serve to amplify the consciousness in an educated and informed sector of the Latin American public opinion, and at the time was capitalized by some of the indigenous organizations and its members. Despite the ideological force and media success of consecutive governments, the support received from the imperial power, in weapons as well as Cold War ideology, upheld the invisibility of the various modes of systemic aggression and violence against the indigenous nations and communities. From the massacres of 1964–65 in the central jungle of Peru at the hands of the police and armed forces, then later in *Madre de Dios*, to the massacres of *Senderismo/Tupamarismo*, and of the armed forces themselves, during the tragic decade, ending with the thousands of *Ashanínka* and *Nomatsiguenga* refugees and other Amazon peoples forced to abandon their territories and exile themselves into absolute poverty, the recent history of the Amazon conveys the irrefutable genocidal political structure of the Peruvian state in relation to the indigenous communities of the jungle. There is only one exception to this genocidal political structure and, as the exception precisely confirms the norm, it makes reference to the few years period of the *Velasquista* Revolution (President General Juan Velasco Alvarado) that offered a political gasp of fresh air and legality toward the indigenous peoples and organizations of the Amazon so that they could adjust themselves to answer the forthcoming neo-liberal market lethal attacks.

The Political Economy of Genocide

Why was Peru's *Velasquismo* an exception in the 1960s? How can we explain that even within the "revolutionary" framework and popularity of the military revolution of Velasco Alvarado, the violent expansion into Amazonian lands decreased, and even bore achievements, in the consolidation of supportive legislative measures answering to indigenous territorial complaints? The explanation of this apparent contradiction must be found within the structural distortion of the Peruvian economy of the 1960s and 1970s in relation to the liberal economic reforms that were being produced in the United States and England, which culminated strengthening the University of Chicago School of Economics, something that gave great momentum to neo-classical economy with the name and program that would be called neoliberalism. In short, this revised version of bourgeois economy, proposed less government and state intervention in the market economy, de-regulating the so called "invisible hand of the market" in production, circulation and consumption. This act of blind faith in the "free market" demands that the role of traditional government or state ceases to intervene and adjust dysfunctions and socio-economic inequalities. Social welfare becomes anathema. The law of competition and efficiency regulates the institution of the market. Prime Minister Margaret Thatcher in the United Kingdom and President Ronald Reagan in the United States were the forbearers of this radical — even extreme — liberal economy. In the 1970s, Peru passed with impunity these kinds of neoliberal reforms while Chile's military dictatorship implemented them at gunpoint, with disappearances and tortures enabling the shameful enrichment of the oligarchy, all while relegating workers and peasants to poverty and further marginalizing the Mapuche

Indians by dispossessing them of all their precious resources. In Perú, with *Velasquismo*, for a change, a very important part of the political leadership and a small circle of military leaders believed — without saying it aloud — in social democracy adapted to meet the needs of city and country. Above all, it was a state with regulatory powers coupled with an ethical commitment to social justice. As it is known, beginning with the coup d'état that brought Morales Bermúdez, the neoliberal gospel once again took control of the state and the government without achieving, however, the total cancellation of the structural reforms established under *Velasquismo*.

Accumulation by Dispossession and Ethnogenocide

What does all of this have to do with the indigenous peoples of the Amazonia? It is here, that one must return to the analytical proposition of David Harvey and his thesis, that in this phase of global neo-liberal capitalism the process of accumulation of capital begins with the dispossession or pillage of resources, labor power, even money that is under the relative control of some classes, groups, or, in the case of the Amazon, of the indigenous ethnic nationalities. The same history of the national expansion of the borders of agro-forestry and now mining/petroleum extraction, explains why the dispossession must take place on the territories of the indigenous peoples. Where else can capital go to pillage and loot? Where there are still standing forests, waters, richness of life, minerals, and unfortunately, petroleum, if not in indigenous Amazonia? This is the country of dozens of indigenous ethnic entities, of indigenous territories, the country that the official Peru of the oligarchy in power considers to be "internal colonies" that can be invaded, occupied militarily, subjugated, conquered, and taken. It was President Fernando Belaunde Terry who coined the phrase worthy of Pizarro: "The conquest of Peru by the Peruvians," is the title of his book. And now this concept of cheap imperialism has undergone the metamorphosis of the "gardener's dog" (it does not eat, but neither lets others eat) in a boisterous presidential display of "profound social-historical analysis." (Also see Devin Beaulieu's contribution in this volume, chapter 11.)

There is a need to dress up with a legal apparatus the dispossession and death by hunger of plundered peoples so that urban public opinion and multilateral financial institutions could digest it. On it, Peruvian post-democracy has learned a lot from U.S. political practices and teachings. Every take over, looting, territorial occupation, illegal appropriation of goods and, finally, all armed violence has to be covered by one or more laws. There is no need to go back to Marx to realize that laws — including those approved and shelved paperwork by diligently, elected by the people, congressmen — constitute a programmatic plan to execute looting with an appearance of civility. This is to say, that it is important for neoliberal governments to keep a certain style of public behavior, so that they distance themselves from drug traffickers and from the stereotypical dictator of banana republics. Nevertheless, the end results are the same: indigenous peoples are dispossessed of their lands that they have occupied for millennia, and of their resources that have been nurtured in a productive way from times immemorial. Who has domesticated the totality of edible and useful plants in Andean Amazonia and the lower jungle? Who has created and recreated the civilized landscape of Amazonian forests that through millennia sustained the lives of millions — not thousands, but millions — of individuals? They were certainly not Europeans, neither were they Peruvian *criollos* of the oligarchic republic of the nineteenth century or contemporary ones, who are rather experts in forest depletion, animal looting, water pollution and the constant abuse of Amazonian peoples.

The accumulation by dispossession was imposed by the government of Alan García and other agents of the financial-business oligarchy as a dominant, extractive economic mode in the Amazon. The victims of this ancient and renewed system of robbery do not only include the indigenous, riverine, and settled poor communities, but also the dense network of social-biotic relationship that allows a renewal of resources that enrich the landscape in its productive diversity and its regenerative capacity. At the end of this criminal mode of extraction and accumulation, only deserts remain, as do unproductive savannahs, entirely disappeared communities, or communities overtly weakened and unable to reconstitute themselves as agents of their own history. Even a superficial reading of the biotic-cultural history of the Amazon will give us clear indication that these will be the final results of the pillage. The tragic irony of the companies of the liberal-bourgeois economy is the scandalous accumulation of wealth in a few hands, national and transnational, that ends up fading in a speculative financial and banking world, fundamentally corrupted by an incestuous reinvestment of the profits to reproduce itself as a class, without leaving any wealth to be redistributed, even partially, toward the bottom, toward the looted people, or toward productive projects with moral objectives and social justice.

What To Do?

Since the 1970s, Amazonian indigenous communities jumped onto the national social, political, and cultural stage in unstoppable ways. The indigenous communities of the Amazon have given themselves over to Peru, a Peru that has been unconscious and therefore, tricked into trinket consumerism, hopelessness, repeated political party disappointments, and a corrupt system. The organized Amazonian indigenous communities are permanent testimony of the Peru that is tormented by the horrors of the outrageous and useless decade of violence. The great and monumental lesson that the Amazonian indigenous communities offer is the need to organize and reorganize, to do without political parties to not be lured by political ideologies and practices generated outside of indigenous historical-cultural trajectories, and to not be deceived by the rhetoric of development and modernization, that during some decades has been the disguise of the global capitalist project.

While the working class is being divided and dismantled within a model of fragmented industrial production that challenges and criminalizes traditional forms of union protest; while unemployment and urban underemployment grows rapidly; and while the indigenous and racially mixed peasantry have been cornered to venture modes of low profitability in an unfavorable competitive market, set against the agro-industries of high technology and concentration of capital; the indigenous towns of the Amazon, on the other hand, have managed to maintain a certain degree of autonomy and disassociation from the capitalist market in the reconstruction and reinforcement of their social economies (mixed economies of subsistence). It is obvious that poverty continues subsisting, and that it is accentuated in those regions where the loss of resources and environmental destruction puts the community in the disjuncture of selling its labor force to extractive businesses, or having to emigrate circularly to other zones. However, the native communities to a large extent have known how to avoid the total dissolution of the ethnic/communal entity by abandonment or apathy, tending instead to reconstruct themselves socially and culturally even amid very unfavorable conditions. This is the case of the *Asháninkas* communities of the Ene River, martyred, deported, and displaced by Sendero Luminoso guerillas and the army confronted in a war that led to an ethnic annihilation with equal cruelty and stupidity by the two sides of the conflict. It is impossible not to see, in this case, that the 1980s' frightening violence against the *Asháninkas* had genocidal intentions as a part of two similar fundamentalist versions of a nationalist project of "assimilation through extermination." In spite of everything, the *Asháninkas* live, grow, adapt, and organize. More so, they give us lessons of hope.

There are neither political formulas nor cultural recipes to share with the indigenous peoples that face renovated forms of state oppression and violence, and the politics of genocide hidden under the rhetoric of modernization and development. And if they existed they would be the fruit of the same historical experience of each oppressed people. We are witnesses, usually far away and absent, that share the sufferings and the anger of the Amazon peoples, and we try to untangle the lies, the tricks, and the delinquent project of this old and new colonizing enterprise called development and modernity. The indigenous resistance to oppression and subtle or open extermination organized by the state, and its national and transnational owners, have a long history that goes back to the sixteenth century. Little has to be taught to the *Awajun*, for example, who in the seventeenth century ousted the Spanish from their territories, or the *Ashaninka* who closed the entrances to the central jungle to the Spanish and Peruvians from 1742 until 1848. What can be offered in the ethno-political organizational field to the *Yanesha* who started to organize to defend their territories in 1967? The only thing that we, in solidarity, could offer is the narration of their histories of autonomy and resistance, the summary of centuries of opposition to subjugation, the profound memory of their utopian social and cultural projects that some of us, indiscreet outsiders, came to glimpse and comprehend through the warmth of their hospitality, in spite of it all.

What happened in Bagua is just a very recent addition to a long trajectory of violence, discrimination, and outright negligence and short-sightedness.

Acknowledgments

The following students assisted in translating this essay from the Spanish version: Sydney Campos, Elayne Dowd, Emma C. Gaskill, Leo Kornfield, Fernando Otero, Uriel Quiñones, José Luis Linares, Eric Maciel, Nina Milliken, Rafa Mora, Elizabeth Sánchez, Victor George Sánchez Jr., David A. Sussberg, Jana Whalley, all members of UCSC's seminar on El Area Andina Contemporánea directed by Prof. Guillermo Delgado.

10 Puentes Para Unir los Fragmentos del Peru

Rodrigo Montoya

"Todos somos indígenas", es la pancarta que en las calles de Lima expresa una inesperada solidaridad con la rebelión de Bagua, una sensibilidad que va mucho más allá de la Amazonía y que compromete al Perú como conjunto. Podría indicarnos también que el terreno podría estar ya preparado para construír los puentes que hacen falta para acabar con la estructura colonial de un país fracturado en segmentos distantes y muchas veces opuestos. Alberto Pizango, el dirigente mayor de AIDESEP, optó por el exilio voluntario de 11 meses en Nicaragua para defender su vida y vuelve para responder a las acusaciones, y porque -tal vez- al oír múltiples y diversas voces de solidaridad con él las autoridades judiciales, comprendieron que no debería ir a la cárcel.

En mi artículo, "Con los rostros pintados: tercera rebelión amazónica en Perú, agosto 2008–junio 2009" (Montoya, 2009), se encuentra una crónica de lo ocurrido en Bagua y una visión de conjunto de lo que fue la rebelión, su contexto, sus razones, sus perspectivas y sus desafíos. Una de las tesis principales del texto sostiene que dicha rebelión es un momento muy importante en la historia peruana en la medida en que define un antes y un después. Después de Bagua, el Perú ya no puede ser visto como un país sólo de criollos puesto que los componentes andinos y amazónicos, excluídos por la república, salen de la exclusión y del mundo subterráneo para exigir sus derechos políticos y reclamar la cuota de poder que les corresponde.

En este ensayo, propongo una reflexión sobre dos momentos de la historia política peruana. En el primero, el ciclo de luchas por la tierra entre 1888 y 1980 impuso en el escenario a los campesinos y a Hugo Blanco como el líder más importante. El segundo podría haber comenzado con la rebelión de Bagua como punto más alto del movimiento indígena amazónico-andino del país. Alberto Pizango parece ser el Hugo Blanco de este momento y hay razones suficientes para suponer que podría convertirse en un dirigente de alcance nacional y no sólo amazónico. En ambos casos, la realidad peruana contradijo y contradice la tesis ortodoxa de la izquierda marxista europea de que la clase obrera sería motor y dirección de la revolución socialista, y que podría después aliarse con los campesinos considerados como fuerza social subalterna. El surgimiento del movimiento político indígena ha sido una sorpresa para lo que queda de la izquierda después de la caída del muro de Berlín y del naufragio de la Unión Soviética.[1] La mayor parte de quienes son aún líderes e intelectuales del espacio de izquierda, no están en condiciones de entender la potencialidad política de ese movimiento tanto para los indígenas como para ellos, ellas, y el país entero, con sus contradicciones, debilidades y desafíos.

Hoy, 2010, la única oposición realmente existente en Perú frente a la política de desarrollo capitalista del presidente Alan García Pérez es la del movimiento indígena amazónico y andino comandado por la Asociación Interétnica para el Desarrollo de la Selva Peruana, AIDESEP, la Confederación Nacional de Comunidades Afectadas por la Minería, CONACAMI, por la Coordinadora Andina de Organizaciones Indígenas, CAOI. Frases como 'Nos hacen falta hombres como Alberto Pizango', 'Los amazónicos sí son capaces de pararle el macho a los apristas', son elocuentes. De la realidad que funda esa percepción podría nacer otro bloque político, formado por indígenas amazónicos y andinos, migrantes organizados en las grandes ciudades, apoyados por los campesinos, sindicatos de obreros, organizaciones de mujeres, profesionales, profesores, estudiantes, y movimientos que defienden sus derechos sexuales diferentes. Pero no será fácil. Aún pesan mucho las ataduras con el pasado, las tesis ortodoxas, los egos de dirigentes víctimas de sus espejos, la intolerancia para no aceptar a los otros como son, el euro-centrismo, y el viejo racismo estructural que anida en mucha gente que vive la ilusión de no ser ni sentirse racista.

Para los dueños de los medios de comunicación y sus periodistas de mayor confianza -que se sienten felices con el bienestar de la economía de las grandes empresas multinacionales y con Alan García Pérez, el movimiento indígena de dentro y fuera del país, particularmente el de los amazónicos y AIDESEP, es una piedra en el zapato, una sorpresa desagradable que no alcanzan a entender. Les encantaría que esos movimientos desapareciesen, aunque han aprendido a aceptar la diversidad cultural — la foto Benetton de los indiecitos bonitos en una carta postal — que ofrece buenos dividendos económicos en áreas como el turismo, la moda y la gastronomía. (Montoya, 2009).[2]

Antes de presentar los dos momentos históricos del Perú contemporáneo, es importante prestar atención a la formación de la izquierda, al papel excepcional de José Carlos Mariátegui y a la ortodoxia marxista europea sobre el papel de la clase obrera como sujeto y dirección de lo que sería la revolución.[3]

Ortodoxia marxista europea sobre la clase obrera, JC Mariátegui y el comienzo de la izquierda peruana

Luego de sus profundos estudios sobre la lógica de funcionamiento del capital y el capitalismo, el proceso histórico de la acumulación de capital y de las tendencias futuras de ese modo de producción, Karl Marx propuso que la clase obrera sería en Europa y en Estados Unidos el nuevo gran sujeto y actor histórico para cambiar el mundo, dejar atrás el capitalismo, construír el socialismo y llegar al comunismo. Con Marx y Friedrich Engels nació la convicción de considerar a la clase obrera como embrión o portadora de otro mundo para intentar convertir en realidad la utopía de la modernidad entendida como sinónimo de justicia y de libertad. Esta visión europea fue llevada al mundo entero y asumida dentro de la izquierda en general como un dogma, una verdad que no necesitaba ser probada. Como demográficamente los obreros no eran mayoría en los países europeos y Estados Unidos, y menos en los del llamado tercer mundo, la clase proletaria fue considerada como motor, dirección y fuerza principal de una alianza de fuerzas entre las que destacó el campesinado, seguido por la llamada "pequeña burguesía", las capas medias e intelectuales. En la segunda mitad del siglo XIX, primeros tiempos de la izquierda marxista europea, la preeminencia de la clase obrera sobre el resto de clases y grupos sociales, parecía inobjetable. El propio Marx comparó a los campesinos con "un saco de papas", queriendo decir que no tenían potencialidad alguna para ofrecer otro tipo o modelo de sociedad. Pero la realidad mostró que en Rusia y China los campesinos debían ser tomados en cuenta tanto por su peso demográfico como por la importancia de sus luchas por la tierra y contra los señores de la tierra. Entonces fue pertinente subir de categoría a los campesinos, quienes pasaron a ser un "aliado principal" de lo que se llamó "Alianza de la clase obrera y el campesinado".

En países del llamado tercer mundo como Perú, la clase obrera se formó en la segunda mitad del siglo XIX, en bolsones principalmente textiles (Vitarte), agrarios (complejos cañeros); luego, en las minas y en algunas ramas industriales, sin haber sido nunca demográficamente grande y, menos, mayoritaria. Luego de la caída del muro de Berlín, la desintegración de la Unión Soviética y los países socialistas de Europa del Este, el Consenso de Washington y la globalización, los capitales se fueron libres por el mundo buscando regiones y lugares de más bajos salarios y menos derechos para obtener ganancias mayores. En contraste, los trabajadores encuentran muros de fierro, cemento y vergüenza para no entrar a Estados Unidos, (Muro de San Diego en la frontera con México, muro en Israel para que los palestinos no pasen a trabajar dentro de su propio suelo, muros de alambre en Melilla y Ceuta para que los africanos no pasen a España y Europa, muro en Irlanda para que los protestantes del norte no pasen al sur, muros interiores en Lima, Buenos Aires o Río de Janeiro para separar a los ricos de los pobres). En esta ofensiva capitalista de gran envergadura, los obreros del mundo pierden empleos, estabilidad y seguridad social, y buena parte de sus derechos sindicales. Por ese camino su "reducción política" es inevitable.

Desde tiempos de la Central General de Trabajadores del Perú (CGTP) del Partido Socialista de Mariátegui (1928) hasta hoy, nunca la clase obrera peruana fue demográfica y políticamente mayoritaria en el país. Los partidos de la izquierda que siguieron las líneas internacionales del partido comunista soviético, y del partido comunista chino, continuaron dentro del pensamiento europeo de la clase obrera como sujeto, motor y dirección, y de la necesidad de una alianza con el campesinado. Por su parte, las tendencias de la cuarta internacional trotskista, se mantuvieron firmes en sus postulados eminentemente obreros. La aparición de Hugo Blanco como un líder trotskista a la cabeza de un movimiento campesino en el Perú fue, seguramente, un hecho que ninguna de las tendencias de la cuarta internacional esperaba.

Sobre ese fondo intelectual y teórico se crearon y florecieron organizaciones sindicales de obreros y organizaciones políticas igualmente de obreros e intelectuales. Según esa lógica, el resto de clases y segmentos sociales tendrían que aliarse a los obreros y aceptar la dirección de éstos. Este esquema eminentemente moderno y europeo fue rápidamente adoptado en América Latina, y seguido con entusiasmo en países de fuerte composición obrera como Argentina y de pequeña y débil presencia obrera como Perú. Con estas y otras ideas volvió Mariátegui de Europa en 1923. Pero tuvo la suficiente lucidez para observar que tres cuartas partes de la población peruana eran indígenas. Inició luego un proceso de acercamiento intelectual, político, amical y afectivo con decenas de líderes andinos que llegaban a Lima con sus memoriales de protesta contra los abusos que sufrían de los gamonales y políticos criollos y, también, con decenas de intelectuales y artistas que defendían la causa indígena.[4]

De esos contactos brotaron textos diversos para la revista Amauta y algunas de las ideas centrales para proponer un acercamiento entre el socialismo y los indígenas. En 1928, Mariátegui fundó la Confederación General de Trabajadores del Perú, CGT, y el Partido Socialista Peruano y estuvo entre sus planes fundar también una confederación y un periódico indígenas. Su muerte, en 1930, frustró esos proyectos. Es pertinente citar aquí siete tesis del amauta sobre lo étnico y el socialismo:

> Tesis 1. El pecado original del Perú es haberse formado sin el indio y contra el indio. Optar por el Perú incaico y renunciar al Perú colonial es la clave sociológica de la sociedad peruana. La literatura comienza a ser peruana sólo cuando incluye al componente indígena.
> Tesis 2. El Perú es una nacionalidad en formación, es un concepto por crear: Los indios -tres cuartas partes de la población- constituyen el cimiento de esta nacionalidad en formación. Sólo una insensibilidad moral explica por qué se le ignora.
> Tesis 3. No es suficiente ver los fragmentos del Perú, lo ideal es tener una visión del Perú integral. No tiene sentido ignorar su componente hispánico. A pesar de la carnicería de la conquista y la explotación colonial, la historia del Perú ha sido creada también con aluviones de occidente. El Perú es sólo un fragmento del mundo. La ciencia y la técnica de occidente constituyen una herencia irrenunciable. El descubrimiento de América, en general, y del imperio incaico, en particular, constituyen el comienzo de la modernidad, de la utopía de una sociedad justa: No existe un conflicto real entre la revolución y la tradición.
> Tesis 4. El socialismo no debe ser confundido con el indigenismo. El indigenismo ha sido útil para enjuiciar al gamonalismo y a la feudalidad. El socialismo no es una solución indigenista y por eso filantrópica del problema del indio. La literatura indigenista ha sido hecha por mestizos. La literatura indígena vendrá a su tiempo.
> Tesis 5. En 1927, sólo es posible empezar a plantear el problema indígena. Resolverlo es un asunto del futuro. El punto de partida para la solución del problema indígena es plantearlo como un problema esencialmente económico y social. Los gestores de la solución de los problemas de los indios deben ser los indios mismos.
> Tesis 6. El socialismo en América Latina no debe ser calco o copia sino una creación heroica. El socialismo es de origen europeo pero existe en la tradición americana. La comunidad campesina puede convertirse en la célula del Estado socialista moderno y la tradición de solidaridad campesina es fundamental. Pero el socialismo no significa volver al socialismo inca.
> Tesis 7. La esperanza indígena es absolutamente revolucionaria. Hay una consanguinidad del movimiento indigenista con las corrientes revolucionarias del mundo. Para que el socialismo sea socialismo y peruano debe solidarizarse con las reivindicaciones indígenas. Es necesario darle a la lucha indígena un carácter de lucha de clases. (Montoya, 1990)

Estas tesis debieron haber sido el punto de partida para pensar e imaginar las posibles vías para el socialismo peruano. Si Mariátegui hubiera seguido fielmente la ortodoxia marxista europea, habría ignorado la llamada cuestión indígena. Felizmente, tuvo una mente abierta y una convicción profunda de que en esos años sólo era posible poner los primeros cimientos de un largo proceso.[5] Ochenta años después, con los sucesos de Bagua y lo que ocurre en el mundo andino, esas ideas mantienen su vigencia pero los tiempos cambiaron: lo que queda de la izquierda no tiene el atractivo que logró tener fugazmente con las luchas por la tierra. Hoy, la esperanza se encuentra en los pueblos amazónicos, andinos y costeños.

Tomas de Tierras, 1960–1980, Hugo Blanco: Oportunidad Perdida

En tiempos de Mariátegui, el partido socialista no participó en elecciones políticas, ni organizó insurrección o guerrilla alguna. El amauta consagró sus siete últimos años de vida a colocar los primeros cimientos de lo que los socialistas debían hacer. Su muerte, la conversión del partido socialista en partido comunista, y los regímenes dictatoriales del país entre 1930 y 1956 cerraron esa etapa. Fueron las luchas por la tierra entre 1957 y 1970 las que situaron a los campesinos como actores centrales de la política peruana, al margen del partido comunista, y las que dieron lugar al surgimiento de nuevas organizaciones de la izquierda. Cuando Hugo Blanco dirigía la toma de tierras en Cusco, en Lima Héctor Béjar y sus compañeros pensaron que esa ola política nueva debía ser apoyada con una organización armada, siguiendo el ejemplo guerrillero cubano. En 1963 fue creado el Ejército de Liberación Nacional, ELN. El poeta Javier Heraud fue acribillado en el Río Madre de Dios en ese mismo año, cuando desde Bolivia intentaba llegar al Cusco junto con un grupo de compañeros preparados para la lucha militar. Ya era tarde. Blanco había caído preso. Luego, la columna guerrillera de Béjar en las selvas de Ayacucho fue igualmente vencida. Lo que me interesa resaltar aquí es la sensibilidad de Héctor Béjar y sus compañeros para oír el mensaje de las tomas de tierras como embrión para crear una fuerza nacional capaz de cambiar el país.[6]

Paralelamente al ELN, otro grupo creó el Movimiento de Izquierda Revolucionaria, MIR, dirigido por Luis de la Puente Uceda que en 1965 tuvo un destacamento guerrillero en la sierra central (Andamarca) y otro en Mesa Pelada, Cusco. Su dirección estaba convencida de que la revolución peruana se haría a través de la lucha armada. El MIR fue igualmente derrotado. En 1968, fue creada una nueva organización de izquierda con el nombre de "Vanguardia Revolucionaria" con el explícito propósito de retomar el camino dejado por Mariátegui, resumido en la consigna "Ni calco ni copia, creación heroica" y volver a colocar en el escenario político a la clase obrera como motor y dirección. Ese mismo año, el gobierno del general Juan Velasco Alvarado promulgó el 24 de junio la ley de Reforma Agraria, la más radical del continente. El estaba convencido que si no se resolvía el problema de la tierra, volverían a aparecer otros Hugo Blanco, Héctor Béjar y Luis de la Puente Uceda y nuevas guerrillas. La reforma agraria de Velasco sumada a las de 1964, 1963, y a la decretada por la Federación Departamental de Campesinos del Cusco, DFCC, en 1962, transfirieron más de diez millones de hectáreas y acabaron con la clase de gamonales y sus siervos, devolvieron a las comunidades gran parte de sus tierras apropiadas por los hacendados y expulsaron de los complejos agroindustriales a la burguesía agraria.

La Central General de Trabajadores del Perú, CGTP, comandó en 1977 el paro nacional más fuerte de la historia peruana que sirvió para que la dictadura militar del general Francisco Morales Bermúdez (1975–1980) se viese obligada a convocar a una asamblea constituyente y abrir un nuevo período de elecciones políticas. En esas elecciones, Hugo Blanco obtuvo la primera votación entre los candidatos de la izquierda y luego de haber sido representante en esa asamblea fue promovido como cabeza de lo que sería la Alianza Revolucionaria de Izquierda, ARI, "sí" en lengua Quechua. Comenzó un proceso de discusión entre todas las organizaciones de la izquierda. Los partidos marxistas-leninistas-maoístas dijeron que no apoyarían nunca a un trotskista, los responsables del partido comunista tomaron la misma posición, el abogado Alfonso Barrantes, estaba convencido de que nunca iría detrás de un trotskista. Además del partido de los Trabajadores, PRT, y otras organizaciones trotskistas, sólo Vanguardia Revolucionaria y el Movimiento Revolucionario Socialista mantuvieron su apoyo a Blanco. Luego de consultas internas y externas, el PRT convencido de su "segura victoria", sin alianzas con los "revisionistas", los maoístas, y el independiente Alfonso Barrantes, los camaradas trotskistas convencieron a Blanco argumentando que él era una figura mundial, y que no necesitaba de aliados. Finalmente, el PRT decidió retirarse de ARI y presentarse solo a las elecciones generales de 1980. Fue lamentable el fracaso nacional, aunque Blanco fue elegido diputado. Quedó la necesidad de unidad y la alianza de Izquierda Unida, IU, alrededor del abogado Alfonso Barrantes. El paso de Hugo Blanco por el congreso no dejó huella alguna y sirvió para prepararle el camino al exilio voluntario y desaparecer de la escena política por varios años. Después de una larga experiencia con los zapatistas en México, volvió a fines de los años noventa con una postura autocrítica y una decisión de apoyar a los movimientos indígenas, dejando de ser el líder de otros tiempos. Más allá de sus cambios políticos, Hugo Blanco no dejó de ser un hombre siempre ligado al pueblo, y vivir en condiciones muy austeras.

Cuando Alfonso Barrantes fue alcalde de Lima, el aprista Alan García fue elegido en 1985 presidente de la República. En ese momento Sendero Luminoso (SL) ya había irrumpido en el escenario como una fuerza política dispuesta a hacer estallar el precario sistema democrático del país, del que Izquierda Unida formaba parte. IU no pudo resistir las presiones desde dentro. Su silencio frente al lamentable primer gobierno aprista por la amistad con Alan García contribuyó a que finalmente IU estallase, como antes había estallado ARI. De la pequeña unidad, sólo quedaron fragmentos dispersos mientras en la Amazonía y en los Andes aparecía primero una organización étnica y más tarde un movimiento político indígena y Alberto Pizango como una enorme sorpresa, del mismo modo que las tomas de tierras de 1960.

Cómo explicar la ruptura de la ARI y el estallido de la IU, es una pregunta fundamental para entender la política peruana. Un listado de **principios no escritos** de las organizaciones políticas y sindicales, tomado de la moderna política europea de izquierda y/o recreados y multiplicados regional y localmente pueden ser una pista para una respuesta: 1. El poder es lo primero; 2. Dentro de la organización hay siempre dos líneas, una correcta entendida como revolucionaria y otra incorrecta considerada como contra revolucionaria; 3. Hay una gran confrontación entre una línea reformista y otra revolucionaria; 4. No hay adversarios, hay enemigos; 5. Las contradicciones se resuelven con la expulsión de los enemigos; 6. Primero muerto antes que aliarse con un enemigo; 7. Alianzas, sí, siempre y cuando se tenga el pleno control de los aliados; 8. Un dirigente es un caudillo, una persona indispensable, un hombre o mujer que consagra su vida a la revolución o a la organización; 9. Sólo se mencionan los intereses de clase, los otros, particularmente los intereses individuales, no son tomados en cuenta. La vanidad y el ego de cada uno(a) no merecen atención ni crítica de fondo; 10. Dividir para alcanzar, ejercer, no perder, o recuperar el poder es un recurso táctico estratégico; 11. El partido debe tener el control de los sindicatos.

Tales son algo así como 'el virus del poder', que las organizaciones sindicales y políticas adoptan pero que no siempre desarrollan. Por eso, hay muchas personas que no comparten esos principios no escritos, se mantienen críticos, resisten, y muchas veces renuncian o simplemente se van. En el examen crítico y autocrítico de cada organización política y sindical podría encontrarse la verdadera magnitud de esas concepciones para hacer política.

Junto con los principios no escritos, los principios declarados formalmente, defienden la unidad, la voluntad del servicio al pueblo, la diferencia entre partidos y sindicatos, la democracia interna, el centralismo democrático, el respeto de los otros, la fraternidad y hermandad. Resulta inevitable una contradicción entre el decir y el hacer. Por eso, parecen inevitables el desconsuelo, la amargura y la falta de esperanza.

Cada organización política puso sus intereses propios por encima de la unidad que se requería para capitalizar positivamente la potencialidad contenida en las tomas de tierras y la solidaridad que éstas provocaban en la población. Es visible la contradicción entre los intereses de una línea internacional y las necesidades nacionales. Dentro de la lógica de construcción del socialismo defendida por la Unión de Repúblicas Socialistas Soviéticas, URSS, correspondía a los partidos comunistas del resto del mundo velar por la URSS como madre del socialismo, ser guardianes de sus éxitos y no ponerla en peligro, embarcándose en proyectos considerados como aventureros. Lo mismo ocurría entre las organizaciones maoístas y los partidos seguidores de su línea.[7]

Bagua 2009 ¿Un Nuevo Momento? Alberto Pizango

El movimiento amazónico tiene una historia de cuarenta años. Aparece con nuevos rostros, nuevos nombres y apellidos, del color de la tierra, con reivindicaciones que poco o nada tienen que ver con el pasado político del país, con prácticas igualmente originales y con tesis políticas que no se limitan a la Amazonía sino que comprometen al conjunto del país, los países andinos, América latina y el mundo entero. Un simple listado de las nuevas reivindicaciones que los movimientos políticos indígenas han elaborado hasta hoy en Ecuador, Bolivia, México, en la Amazonía, en los Andes, es el siguiente: 1. la tierra es nuestra madre; 2. el territorio es un gran espacio entre ríos, lagunas y quebradas para la vida, pesca, agricultura, espiritualidad y universo mítico, dentro del Estado peruano y no en contra; 3. la defensa de la cultura; 4. la defensa de la lengua; 5. defensa de la identidad como ejercicio de sus derechos de ser diferentes; 6. su ciudadanía étnica para afirmar que son peruanos e indígenas de cada uno de los pueblos amazónicos; 7. su derecho a la libre determinación; 8. su derecho al auto gobierno; 9. por el deber de mandar obedeciendo; 10. por construír una sociedad sin la concepción occidental del poder; 11. por su defensa de los derechos colectivos de los pueblos; 12. por el respeto a las mujeres; 13. por el respeto a los pueblos y personas; 14. por el respeto a la dignidad de los pueblos y personas; 15. contra la exclusión; 16. contra la indiferencia; el abandono y el olvido; 17. por la defensa de la biodiversidad; 18. por la defensa de los recursos de la Amazonía como bienes colectivos de todas las peruanas y peruanos; 19. por la defensa de los saberes indígenas acumulados en miles de años de organización del espacio amazónico; 20. por una educación intercultural bilingüe; 21. por la defensa de la espiritualidad propia de los pueblos; 22. por la creación de un Estado plurinacional; y 23. por el "buen vivir", que podría ser una alternativa al llamado desarrollo que en occidente busca la riqueza a cualquier precio como condición para vivir mejor.

Entre las nuevas prácticas habría que señalar, la creación de organizaciones étnicas, diferentes a las ya conocidas de carácter sindical o político, el esfuerzo de construcción de sus organizaciones desde abajo, su modo de hacer política cantando, bailando, con alegría, y hasta elegantemente vestidos.

Merece atención especial la propuesta de los pueblos amazónicos para defender los recursos naturales (petróleo, gas, oro, madera, biodiversidad, saberes indígenas, particularmente médicos) como bienes públicos de todos y no como propiedad de unas cuantas empresas multinacionales en directa alianza con gobiernos. Su defensa de la Amazonía la entienden como defensa del planeta, de toda la humanidad.

Todo lo que acabo de presentar, de modo tan breve, no habría sido posible sin el notable esfuerzo de los intelectuales indígenas bilingües y biculturales que luego de haber aprendido a leer y escribir, y haber obtenido una formación en colegios y universidades, tienen un juicio crítico sobre su propia realidad, el país, América Latina y el mundo, y van construyendo un proyecto político y organizando fuerzas propias y alianzas para realizarlo. En su conjunto representa una laternativa política muy distinta a aquella propuesta por los partidos políticos desde 1960 hasta ahora.

La rebelión de Bagua se produjo dentro de un contexto internacional marcado por el caso particular de Bolivia, país en el que ya está vigente una nueva constitución que por primera vez en su historia toma en serio la representación de los pueblos indígenas, propone la tarea de construír un estado plurinacional y sostiene el Allin Kawsay o buen vivir como alternativa frente al desarrollo capitalista de la acumulación de riqueza como condición para vivir mejor. En la misma línea, pero fuera del estado y el gobierno, el movimiento indígena ecuatoriano mantiene su orientación básica y trata de resolver los problemas creados por los gravísimos errores de haberse aliado con un caudillo militar, el coronel Gutiérrez, y de no haber sabido controlar a una parte de sus representantes en dicho gobierno que defendieron sus intereses personales antes que los del movimiento indígena ecuatoriano. (Ver capitulo de Cooper en este volumen).

Problemas y Perspectivas

No todo es color de rosa, ni todo lo nuevo que aparece en el horizonte viene puro y transparente. Lo nuevo viene entre brumas, de modo casi imperceptible. Se requiere de una mirada atenta y profunda para detectar los embriones de cambio que la realidad contiene sin que a veces sus propios actores se den cuenta de la importancia que tienen. Presento a continuación algunos de los problemas más importantes:

1. En los últimos treinta años la afirmación étnica y política ha sido posible al mismo tiempo que se intensifican los procesos de desindigenización en la medida en que el modo de vida itinerante en segmentos pequeños de linajes derivados de ancestros comunes se modifican seriamente desde 1974, cuando en el gobierno del general Juan Velasco Alvarado se fundan los centros poblados permanentes de las comunidades nativas, alrededor de una escuela y un campo de fútbol. Este nuevo fenómeno urbano ha producido la fuga de los animales bosque adentro y ha producido una sustantiva disminución de la pesca disponible en ríos, lagunas y quebradas cercanos

a los centros poblados. La categoría pobreza aparece con ese cambio y con la conversión de nativos en campesinos que producen para el mercado y por eso son vendedores baratos y compradores caros.

2. Como consecuencia de la firme decisión de los padres de familia para enviar a sus hijos a escuelas bilingües y monolingües, crece el bilinguismo castellano-lengua originaria, disminuye el monolingüismo en lengua originaria, aparecen niños en edad escolar cuya lengua materna es el castellano y, como consecuencia, los padres de familia piden que en las escuelas del Proyecto de Educación Bilingüe de Iquitos, por ejemplo, se les enseñe su lengua propia como si fuera una segunda lengua.

3. La soledad indígena sigue siendo un problema debido a la ausencia de aliados firmes y confiables. El caso del grave error de los indígenas ecuatorianos al aliarse con el coronel Gutiérrez es el más ilustrativo. La soledad fue y sigue siendo aún un problema de los campesinos y luego de las tomas de tierras. Una alianza entre los pueblos amazónicos, andinos y costeños se encuentra aún en un momento inicial de construcción. No es fácil dejar atrás viejos conflictos del pasado derivados de los abusos causados por los expropiadores-depredadores caucheros, petroleros, buscadores de oro, colonos de los Andes expulsados por la pobreza inducidos a sobrevivir en la Amazonía, y también de los conquistadores católicos y evangélicos de almas indígenas. El caso boliviano es una excepción.

4. La nueva condición de pobres es una limitación profunda para que los pueblos indígenas de la Amazonía tengan acceso a una buena y generalizada educación para ejercer sus derechos políticos en mejores condiciones. Lo mismo ocurre con los pueblos andinos.

5. Aparecen algunos síntomas de división propios de las organizaciones sindicales, políticas e institucionales como las ONGs.

6. Los pueblos indígenas en general carecen de recursos para financiar sus posibilidades de desarrollo y los gastos de sus organizaciones; requieren por eso de la solidaridad internacional que debe ser tratada con extremo cuidado.

7. Se observa una distancia entre el nivel alcanzado por algunas direcciones indígenas y sus bases.

Las organizaciones indígenas amazónicas tienen que hacer frente a numerosos desafíos, particularmente a uno que me parece muy importante: se trata del aprendizaje que deben hacer de formas de organización que no son suyas y que forman parte de la tradición política occidental compartida por los partidos políticos tanto de izquierda como de derecha. La organización social conocida y vivida por los pueblos amazónicos como eje de su sistema de parentesco es la *maloka*, que corresponde a la representación que el jefe de un linaje o de un segmento de linaje asume del conjunto de personas que forman esos segmentos. La *maloka* se confunde también con una casa grande, en la que son recibidos ritualmente todos los miembros en ocasiones especiales, que es diferente a la vivienda de cada grupo familiar y que, sólo por extensión, podría llamarse comunal. Por eso, la noción misma de "comunidad nativa" es una forma de organización estrictamente ajena a la tradición amazónica, pero que fue indispensable para que los pueblos amazónicos fuesen reconocidos por el estado y para ejercer los derechos y deberes que esa pertenencia impone. En el caso de las comunidades campesinas de los Andes, se trata de una organización híbrida formada por el antiguo ayllu inca como conjunto de familias, como unidad básica de la sociedad, redefinida dentro de una estructura formal que correspondía a los municipios españoles, con alcaldes mayores y menores, alguaciles, regidores, campos. Una vara de chonta era, y sigue siendo, aún el símbolo de un sistema de autoridades que nació con las llamadas "reducciones de indios" a partir del virreinato de Toledo en 1569). Con el reconocimiento legal de esas comunidades en 1920 y luego de los nuevos estatutos comunales posteriores a la reforma agraria velasquista de 1969, se impusieron los nuevos cargos de presidentes, personeros y consejeros (de administración y vigilancia). Finalmente, se volvió a las Junta comunales de presidentes, secretarios, tesoreros, y vocales. Los nombres variaron, pero no la estructura de los cargos, que se eligen por un año, periódica y rotativamente, sin remuneración alguna. Si en algún lugar del país tiene sentido pleno hablar de una vocación de servicio de los dirigentes es en las comunidades campesinas, en las que los comuneros que ocupan el alto puesto de Presidentes entregan cuatro o cinco años de sus vidas al servicio de la comunidad a cambio del reconocimiento y prestigio que son importantes. Cuando este sistema fue inventado en el siglo XVI, el ideal de un sistema político democrático, no existía. Y si continúa luego de muchos siglos, como la institución peruana más antigua, es porque se trata de una democracia comunal en la que los abusos del sistema político global llamado democrático no tienen lugar.

El sindicato, el partido político y la ONG, son formas de organización ajenas a las comunidades campesinas y nativas. Adoptarlas como propias significa cargar con el activo y el pasivo de esas organizaciones y vivir el grave conflicto entre respetar las reglas que formalmente se declaran o seguir los principios no escritos que son parte del virus del poder. La tradición amazónica y andina de vivir en comunidades en las que una de las preocupaciones principales es buscar el bienestar de todos los miembros, será el recurso mayor para resolver los problemas que deriven de la aparición de diversos síntomas del virus del poder.

Una cuestión central en la coyuntura de hoy es la tentación electoral. Gran parte de las organizaciones de la izquierda se atomizaron por haber caído en conflictos fratricidas, causados también por disputas de sus dirigentes por una curul. Si estos se preocuparon tanto por tener una silla en el congreso, la tarea de pensar en serio la política para ofrecer un horizonte y una dirección para todo el país quedó como un espacio vacío. Esta es la pequeña gran historia desde 1979 hasta hoy, con algunos congresistas de izquierda que cumplen bodas de plata sin que el movimiento político de base y alternativo haya avanzado.

Una simple disposición en tiempos del fujimorismo para remunerar a los alcaldes y ofrecer dietas a los regidores y la ley que otorgó funciones regionales a los antiguos departamentos, ampliaron notablemente el espacio electoral. Cuando las alcaldías eran ofrecidas a los llamados notables, que no tenían necesidad de recibir un salario, interesaban a muy poca gente. Hoy, miles de personas hacen grandes esfuerzos para ser presidentes y consejeros de las 24 "regiones", alcaldes y regidores en 179 provincias y en 1,747 distritos. Las intenciones profundas tienen mucho que ver con los sueldos que se pagan y, sobre todo, con la posibilidad de administrar mucho dinero. La vocación de servicio existe, seguramente, pero para un número pequeños de candidatas y candidatos. Cuando no se tiene un trabajo y un salario estables, cuando las remuneraciones por el liderazgo temporal acaban, convertirse en dirigentes remunerados por el sistema político es una alternativa demasiado seria pero . . . para ganar elecciones importantes hay que tener dinero, "hay que invertir", en apoyos, aliados, partidos, publicidad en los medios, regalos de bolsas de aceite, leche, azúcar, harina, arroz, polos y gorros con los nombres de los candidatos. Por esta vía, las posibilidades de los dirigentes indígenas y campesinos son muy reducidas.

Lo ideal sería que a partir de la rebelión amazónica se crease un movimiento político de largo plazo en el que confluyan numerosas fuerzas desde abajo y que en el paso a paso del camino vaya formándose un nuevo liderazgo con nuevos rostros, otros apellidos y mejores ideas. Lo importante es el mediano y largo plazo y la tarea de construír a partir de las luchas concretas a nivel local, regional y nacional. Nos hacen falta muchísimos puentes que unan el archipiélago de hoy: puentes entre los pueblos amazónicos, andinos costeños, migrantes de todas partes en los conos de Lima y en los pueblos jóvenes de las ciudades; puentes entre las organizaciones indígenas y las organizaciones de trabajadores, campesinos, maestros, mujeres, estudiantes, periodistas, intelectuales, de grupos que defienden sus derechos de ser sexualmente diferentes; puentes con los migrantes en el exterior, principalmente con aquellos que vuelven a sus pueblos a pasar los cargos que las familias de hoy ya no pueden asumir. Si se construyen esos puentes, la mayoría electoral del país habría sido alcanzada y se puede tener la seguridad de ganar las elecciones nacionales y, además, defender organizada y masivamente esa y otras victorias.

Para construír esos puentes, disponemos ya de algunos pilares. El primero, es el conjunto de 24 reivindicaciones-objetivo propuestos por los movimientos políticos indígenas de los países andinos y de América Latina y presentados líneas arriba en este ensayo. El segundo, es la contribución de teoría política ofrecida por Aníbal Quijano y otros autores sobre la colonialidad del poder, clave para entender la dimensión mundial del patrón de dominación y explotación global del capital en las esferas económica, social, política, cultural, étnico-racial, patriarcal (sexo y género) a partir del comercio colonial con Europa, Asia y África. Paralelamente a esta colonialidad del poder, Quijano & Immanuel Wallerstein centran su atención en los procesos de descolonialización a través de movimientos sociales anti sistémicos que redefinen los conceptos de revolución y democracia en el contexto de una crisis de la civilización capitalista y agotamiento de la modernidad.[8] El tercero, de teoría política como el anterior, podría estar en proceso de gestación al privilegiar el análisis de las relaciones de todas las culturas existentes en cada uno de los países con el poder, particularmente la conversión de la cultura en política que daría lugar a la aparición de un nuevo sujeto político. Mis trabajos últimos apuntan en esa dirección. Son plenamente compatibles la dimensión mundial de la colonialidad del poder y la dimensión nacional de culturas convertidas en sujeto político. Es pertinente recordar que los mejores puentes son aquellos que se trabajan y hacen desde las dos orillas del río. Con ese horizonte, las alianzas son posibles y las soledades podrían quedar atrás.[9]

Para terminar, habría que rescatar de la tradición gramsciana una propuesta que podría ser indispensable: antes de ¨llegar al poder¨ o de ¨ganar el poder¨, la dirección política debe convertirse en autoridad, ser respetada por las ideas y propuestas que ofrece para cambiar la sociedad y por la competencia y honradez personal de sus líderes. Gramsci propuso la noción de hegemonía en el preciso sentido de dirección ofrecida a los sectores sociales por sus organizaciones políticas para formar una vasta alianza política que se convierta en un bloque de poder. En nuestro tiempo, un requisito para ganar esa hegemonía es proponer otro modo de organizar horizontalmente las relaciones sociales, renunciando explícitamente a la metáfora del poder arriba y la base abajo, a la división entre los que mandan y los que obedecen y siguiendo el consejo de los zapatistas para mandar obedeciendo y limarle las patas a la silla del poder para que la silla desaparezca y nadie más trate de sentarse en ella. Es hora también de incorporar dos conceptos claves con los que se crean los consensos en las comunidades indígenas y campesinas: sumar y tejer desde abajo, paso a paso, sin prisa.

Notes

1 El resto de la clase política tampoco tenía idea de lo que pasaba en la Amazonía y en los Andes. La categoría clase política corresponde al conjunto de organizaciones políticas que a pesar de sus diferencias ideológicas comparten una misma práctica de vivir y hacer política.

2 Alejandro Trevisan, un lector de la revista Caretas, escribió el 27 de mayo de 2010: ¨El líder indígena Alberto Pizango debe ser escuchado por todos los peruanos, ya que representa al marginado pueblo de la Amazonía y es una de las personas más coherentes presentes en el actual panorama político. Si Pizango llegara a ser presidente del Perú se comenzaría a saldar la triste deuda que han dejado los quinientos años de conquista y evangelización de los pueblos originarios de América¨. Los responsables de la revista Caretas le responden: ¨Todo el país es testigo que debido a la actitud incoherente o engañosa de Pizango se generaron los luctuosos sucesos en Bagua el 5 de junio de 2009, y ello no obstante la paciente buena voluntad del entonces premier Jehude Simon. Ahora la juez Carmen Arauco Benavente le ha concedido libertad ligada a su comparecencia con impedimento de salida del país. Pasar de ser asociado con el Baguazo a la presidencia es una alucinación de quienes quieren aprovecharlo¨. Los textos muestran la distancia que hay entre el lector Trevisan que tiene un espíritu crítico y los responsables de la revista, tan próxima al gobierno de Alan García, que en este caso preciso no tienen espíritu crítico alguno.

3 En los primeros treinta años del siglo XX los andinos, campesinos Quechuas y Aymaras, aparecieron en Lima protestando. En esos años en la Amazonía terminaba la era del caucho que comenzó en 1880 y produjo una caza de indígenas por las empresas esclavistas como Arana y Fiztcarraldo y sus aliados ingleses. Cerca de sesenta mil indígenas murieron para que los caucheros amasasen sus fortunas. Por las distancias y la casi nula conexión interna no llegaron a Lima delegaciones amazónicas como sí lo hicieron las andinas. En esos tiempos predominaba plenamente el prejuicio contra los llamados ¨chunchos¨ La Amazonía era demasiado lejana y el propio Mariátegui estuvo influído por las ideas europeas del primitivismo atribuido a los "chunchos", habitantes de la Amazonía, considerados como "salvajes". La defensa de los amazónicos hecha por la primera abogada peruana Miguelina Aura Acosta Cárdenas fue mencionada por el Amauta, pero no fue suficiente. Sobre el problema del caucho ver el último libro de Alberto Chirif *Imaginario e Imágenes de la Época del Caucho* (Chirif y Cornejo, 2009). Lo mismo ocurría con los prejuicios racistas contra los negros. Fueron los pueblos andinos los primeros en recibir la atención limeña de artistas indigenistas como José Sabogal y de socialistas como Mariátegui. Casi un siglo después, los amazónicos aparecen con sus propias voces.

4 Sobre la originalidad de Mariátegui ver Aníbal Quijano: *Reencuentro y debate: una introducción a Mariátegui* (Quijano 1981), José Aricó, *Mariátegui y los orígenes del marxismo latinoamericano* (Aricó, 1978), Alberto Flores Galindo, *La agonía de Mariátegui, la polémica con la Komintern* (Flores Galindo, 1991) y Rodrigo Montoya *El problema étnico y el socialismo en tiempos de Mariátegui* (Montoya, 1994). Ver también mi Prólogo a la segunda edición en portugués del libro Siete ensayos de interpretación de la realidad peruana (Montoya, 2008b).

5 Béjar en un homenaje a Javier Heraud — Casa de Mariátegui, en mayo de 2010 — contó que para su grupo de combatientes se trataba de una "lucha armada con minúsculas", ¨de un sincero esfuerzo de los jóvenes por acercarse al pueblo de modo emotivo, sentimental,

no demagógico". Dijo que ninguno pensaba convertirse en héroe, que sólo trataban de ligarse al pueblo para compartir su suerte. Contó también que los jóvenes querían formarse en el marxismo y que cuando el poeta Edgardo Tello, su amigo y camarada, cayó muerto, tenía en uno de los bolsillos de su camisa el texto Manuscritos filosófico-económicos de Marx.

6 En Francia, en mayo de 1968, la clase obrera y el pueblo francés tuvieron el poder cuando la huelga general obligó al general De Gaulle a buscar apoyo en los militares franceses asentados en Berlín. El Partido Comunista francés en vez de seguir adelante y confirmar la victoria, prefirió seguir el consejo del PC de la Unión Soviética, para cambiar la huelga general por las elecciones con el viejo argumento de que las condiciones internacionales no estaban reunidas para un triunfo nacional. En esas elecciones ganó Georges Pompidou el candidato del general De Gaulle y el PC francés sufrió una gravísima derrota de la que hasta ahora no puede recuperarse. Después de la Segunda Guerra Mundial las fuerzas socialistas de Grecia tenían el poder en sus manos pero tuvieron que renunciar porque el acuerdo entre las potencias dejaba a Grecia en el bloque de Europa y a los países de Europa del Este dentro de la órbita de la URSS. Sobre la rebelión francesa en mayo de 1968, ver el artículo "Movimiento Estudiantil y Trabajadores, Francia 1968" (Montoya, 1968).

7 En otro texto (Montoya, 2008) me he referido con más detalle a los problemas que acabo de citar y a otros más que acompañan la emergencia del movimiento político indígena como nuevo actor en el escenario de los países andinos y Latino Americanos.

8 Sobre el tema de la colonialidad del poder, el sistema mundo, la economía y crisis mundial, claves para entender la política, ver los trabajos de Aníbal Quijano: 2000 a, 2000b 1997, 1998, 1994, 1992, 1991, 1988, 1981, 1972, de Immanuel Wallerstein: 2006, 2005, 1999, 1997, 1998, 1997, 1979; de Andrés Gunder Frank:1982, 1972, 1969, 1967 ; de Boaventura de Souza Santos: 2009[a], 2009b, 2008, y 2007; de Enrique Dussel, 2000 y de Edgardo Lander: 2000, 1997.

9 Hablando de puentes, me parece importante citar una propuesta del economista Hugo Cabieses para que en el Perú se construya una alianza de guacamayos, gallitos de las rocas, águilas, cóndores y pelícanos, aves que habitan los más importantes pisos ecológicos del país.

References (Referencias bibliográficas)

Aricó José, selección y prologo. 1978. *Mariátegui y los orígenes del marxismo latinoamericano.* Cuadernos de pasado y presente. Segunda edición, México.

Boaventura de Souza Santos. 2009a. *Una epistemología del sur. La reinvención del conocimiento y la emancipación social.* Buenos Aires: Siglo XXI Editores.

—— 2008. *Pensar el Estado y la sociedad: desafíos actuales.* La Paz: CLASCO, CIDES-UMSA, Muela del Diablo Editores, Comuna.

—— 2007. *La reinvención del Estado y el Estado plurinacional.* Santa Cruz de la Sierra: CENDA, CEJIS, CEDIB, Bolivia.

Chirif, Alberto y Manuel Cornejo. 2009. *Imaginario e imágenes de la Época del Caucho.* Lima: Grupo Internacional de Trabajo sobre Asuntos Indígenas (IWGIA) y el Centro Amazónico de Aplicación Practica (CAAAP).

Dussel, Enrique. 2000. *Transmodernidad e interculturalidad. Interpretación desde la filosofía de la liberación.* Internet, Artículos en Línea, versión pdf.

Flores-Galindo, Alberto. 1891. *La agonía de Mariátegui, la polémica con la Komintern.* Lima: Desco.

—— 1989. Con Ricardo Portocarrero, *Invitación a la vida heroica de José Carlos Mariátegui.* Antología, Instituto de Apoyo Agrario. Lima.

Gunder Frank, André. 1982. *Dinámica de la crisis global* (con Samir Amín, Giovanni Arrighi e Immanuel Wallerstein.

—— 1979. *La acumulación mundial 1492–1789*, Siglo 21.
—— 1969. *Sociología del desarrollo y subdesarrollo de la sociología: el desarrollo del subdesarrollo.*
—— 1967. *Capitalismo y subdesarrollo en América latina.*
Jonathan, Mark. 1994. *Human Biodiversity, Genes, Race and History*, Aldyne de Gruyter, Nueva York.
Lander, Edgardo. 2000. *La colonialidad del saber: Eurocentrismo y ciencias sociales*. Perspectivas Latinoamericanas. Buenos Aires: CLACSO.
—— 1997. *Colonialidad, modernidad, postmodernidad, Anuario Mariateguiano*, vol. IX, no. 9, Lima.
Montoya Rojas, Rodrigo. 2009a. Con los rostros pintados: tercera rebelión amazónica en Perú, agosto 2008–junio 2009. Lima, 92 pp. El texto circula en varias redes de Internet (Servindi, Democracia global) y ha sido publicado por Ediciones Lucha indígena, en noviembre 2009.
—— 2009b. Cuando la cultura se convierte en política, Conferencia inaugural del Congreso de Investigaciones en Antropología. "Retos y perspectivas de la Antropología ante el Perú actual." Universidad San Antonio Abad del Cusco, Cusco, 17–29 de Noviembre.
—— 2008a. Coord. Con Claudia Balarin, Patricia Balbuena, Leonardo Feldman, Raphael Hoetmer, Carlos Infante, Tomas Miranda, Ramon Mujica, Ulpiano Quispe, Alejandro Rey de Castro. *Voces de la tierra: Reflexiones sobre movimientos políticos indígenas en Bolivia, Ecuador, México y Pero.* Lima: Fondo Editorial de la Universidad Nacional Mayor de San Marcos.
—— 2008b. Prologo a la segunda edición en portugués del libro, *sete ensaios de interprecao da realidade peruana*, de José Carlos Mariátegui, Expressao Popular, CLACSO libros, Colecao Pensamiento Social latino americano, traducido por Felipe Lindoso, Sao Paolo. Brasil.
—— 1994. "El Problema Étnico y el Socialismo en Tiempos de Mariátegui," *Anuario Mariateguiano*, Lima VII: 3, pp. 67–82.
—— 1990. Siete Tesis de Mariátegui sobre el Problema Étnico y el socialismo, *Anuario Mariateguiano*, N 2. Lima, pp. 45-70.
—— 1968. "Movimiento Estudiantil y Trabajadores, Francia 1968," *Revista Proceso*, Universidad Nacional Agraria, Lima, pp. 79–88.
Quijano, Anibal. 2000a. "Modernidad y democracia: intereses y conflictos," *Anuario Mariateguiano*, vol. XII, no. 12, Lima.
—— 1998. "La nueva heterogeneidad estructural de América Latina," en Heinz Sonntag. *Nuevos temas, nuevos contenidos*. Caracas: UNESCO Nueva Sociedad.
—— 1997. "Colonialidad del poder, cultura, y conocimiento en América Latina," *Anuario Mariateguiano*, vol. IX, no. 9, Lima.
—— 1994. "Colonialidad del poder, eurocentrismo y América Latina."
—— 1991. "América, el capitalismo y la modernidad nacieron el mismo día," entrevista en *ILLA*, no. 10, Lima, enero 1991.
—— 1992. "Colonialidad y modernidad/racionalidad." *Perú Indígena*, vol. 13, no. 29, Lima.
—— 1988. *Modernidad, identidad y utopía en América Latina*. Lima: Ediciones Sociedad y Política.
—— 1981. *Reencuentro y debate: una introducción a Mariátegui*. Lima: Mosca Azul Editores.
—— 1972. *¿Qué es y que no es el socialismo?* Lima: Ediciones Sociedad y Política, 1972.
Wallerstein, Immanuel. 2006. *Análisis de sistemas-mundo*. Una introducción. Madrid: Siglo XXI Editores.
—— 2005. *La crisis estructural del capitalismo*. México: Editorial Contrahistorias.
—— 1999. *Movimientos antisistémicos*. Con Giovanni Arrighi y Terrence K. Hopkins. Madrid: Akal.
—— 1998. *El moderno sistema mundial III. La segunda era de gran expansión de la economía-mundo capitalista*, 1730-1850. México: Siglo XXI Editores.
—— 1997. *El futuro de la civilización capitalista*. Barcelona: Icaria..

Acknowledgment

Agradezco a mis amigos Alberto Rocha, Hugo Cabieses y José Martínez por sus comentarios.

11 *El Perro del Hortelano*: Producing Activist Film in the Peruvian Amazon

Devin Beaulieu

> What would the Jews do with the Amazon? I don't even want to think about it. Surely, they would plant double in each hectare what nature plants there and be leaders of the world timber market, and furniture, with millions of jobs in the interior. Us? Nothing! And why? Because Congress does not want to approve a law to permit private property. Seventeen million hectares, that for our dear peasant brother have no value — but for an investor, with lots of resources and technology, have an immense value. But someone has invented that these are sacred lands. Investment is the best instrument today to end poverty. Investment is not what creates problems for the environment. Peru will be a petroleum exporting country.
>
> Alan García, President of Perú, Radio Program
> "The Syndrome of the Gardener's Dog"

Introduction

In 2009, the Selva Rica Project, an international arts cooperative, embarked on a full-length narrative feature film, titled *El Perro del Hortelano* (The Gardener's Dog).[1] Rooted in a series of writing and producing workshops with volunteer artists and youth participants, Selva Rica constructed a film about the conflicts that emerge when development meets environmentalism and cultural heritage. Grounded in a mission to promote indigenous protagonism and to create work that reflects the concerns of the place where it is made, the cooperative worked with the Harakambut community of Shintuya to develop a narrative story on the problems of oil exploration by Hunt Oil and Repsol Peru in the Amarakaeri Communal Reserve, within the hydrocarbon concession Lot 76.

We used a radical dialogic method to co-produce the feature film between indigenous and non-indigenous participants. An emergent dialogic process of collective writing, acting, and shooting proved extremely effective for the production of cinema able to critically portray the concrete dilemmas of contemporary Amazonian groups otherwise systematically excluded from self-representation in popular media. The project highlights the possibilities of intercultural engagement through arts and

Figure 11.1 Bora-Huitoto painter, Brus Rubio Churay in the rainforest of Madre de Dios, Peru.

activism, with specific indigenous and environmental context, while speaking to broader national and international questions and politics of Amazonia. The project's achievements have won recognition with multiple international film awards.[2]

El Perro del Hortelano is a feature comedy directed by Renzo Zanelli Baretto about Brus, an indigenous artist, and his efforts to organize his community against an American oil company, Kenny Oil. Searching for answers, Brus joins a local NGO,

Figure 11.2 Ellias "Capi" Aribildo, Brus Rubio, and Wili Corisepa welcome Angie, the American researcher.

finding himself entangled in the subversive work of an American researcher, Angie. As Brus explores the surreal world of volunteers and development experts, he ultimately discovers his own way of bringing strength to his community.

Discovering A Critical Voice

The *El Perro del Hortelano* production took place at the Manu Learning Center and indigenous community of Shintuya near Manu National Park. Manu is a UNESCO world heritage site, and one of the most bio-diverse regions of the world. The area is also home to the Harakmbut indigenous people in the Amarakaeri Communal Reserve. Here, artists from five different Peruvian indigenous communities, including Machiguenga, Harakambut, Bora, Huitoto, and Yanesha created, alongside artists and activists from Peru, Argentina, Spain, USA, Australia, Nicaragua, and Russia, a feature-film about environmental and cultural concerns in the region.

While production began with collaborative workshops on indigenous arts and cultural practices, we quickly changed directions as it became clear that the most pressing concern facing the Harakmbut community of Shintuya was impending oil exploration on their indigenous territory of the Amarakaeri Communal Reserve, encompassing more than four hundred thousand hectares. Several weeks into the project, the director of the conservation center, where we based production, shared with us his frustrated dealings with Hunt Oil and their plans to begin exploration operations and eventual extraction in the area. The basic details of the large-scale operation, affecting an area greater in size to the entire reserve, came as a shock to both local as well as foreign members of our crew. Capi Aribildo, a young man from Shintuya, responded that no one in his community was aware of the details and that

Figure 11.3 Representatives of "Kenny Oil" in Shintuya, played by Kenneth Luna and Rodrigo Carbajal: *"Who wants presents? Who? Who? — Me! Me! – All we have to do is bring presents for their parents, and everything will be easy."*

he feared for his future and ability to live a healthy life. This concrete problem gave the production its impetus and direction, with a critical theme reaching beyond Shintuya and Madre de Dios. Valbina Miguel, a Yanesha community leader, implored the rest of the group at the meeting to take up the issue, as the same events were taking place in the communities in the central Peruvian Amazon.

The Amarakaeri Communal Reserve was created in 2002, following decades of local demands to protect the territory against pervasive logging and mining. In 2006, the subsoil hydrocarbon rights were leased by the Peruvian government under "Lot 76" to the Texas based Hunt Oil and Repsol Peru. Under Alan Garcia's government, 72 percent of the Peruvian Amazon was allotted in multinational hydrocarbon concessions. The concessions are a part of a larger push by the Peruvian state for neoliberal development of the Amazon, tied to free trade agreements with the United States ratified in 2006 and enacted in 2009 (Skeen, 2010). Our production coincided with the start of seismic exploration in the reserve by Hunt Oil, as well as protests and general strikes across the Peruvian Amazon against hydrocarbon concessions and the government of Alan Garcia. These protests led by Peru's indigenous sectors climaxed just weeks following the end of our film shoot on June 5, 2009, with the Bagua massacre, in which Peruvian police killed more than thirty protestors at a roadblock in Amazonas, northern Peru (Renique 2009).

We faced the challenge of making a high quality feature-length film depicting the most important issue facing the Peruvian Amazon during a short, three month shoot, with a budget of less than eight thousand dollars. We turned the camera on ourselves and the unfolding scenes of drama and farce as we worked as artists becoming activists. As Oleg Kheyfets, our enthusiastic Russian theater acting coach, would declare typically each dawn, "What do I need actors for when I have characters like these?!" referring to the motley assembled crew ranging from an ex-Wall Street stock trader, environmental and documentary film makers, Limeño artists, members of the Shintuya community, indigenous artists and activists from across the Peruvian Amazon, myself an aspiring Latin America scholar, and a lost Spanish tourist.

Rather than attempt a documentary project, we cast pretenses to detached objectivity aside and collaboratively produced intentional fictional scenes, and a narrative plot depicting ongoing realities. Often mixing documentary footage with scripted and improvised action, we utilized participants' vernacular expressions and casual interactions through a method of reflective pedagogic dialogue to both inform our audience and the community of Shintuya, with which we worked, of the actual practices and consequence of oil exploration and exploitation on their ancestral lands.

The community of Shintuya received no direct consultation from Hunt Oil prior to the start of seismic exploration on their territory, in violation of the requirements of international indigenous rights regimes, such as International Labor Organization Convention 169. Instead, we discovered a fifty-five-page environmental impact report on the Hunt Oil project by the Peruvian firm, Demus, quietly filed away in the local municipal office of the nearby town of Salvación. "Consultation" in reality consisted of a bureaucratic ritual checklist for Hunt Oil to pass through with the Peruvian government; complete with an ironic "translation" of the text of the executive summary into the oral language of Harakmbut. For "compensation", the then newly elected President of Shintuya received a promise from Hunt Oil of thirty thousand dollars that most community members understood as a bribe.

Organizing our production in response to this context, we scripted the narrative

Figure 11.4 Brus speaking to the community of Shintuya: *"To live within the community is the most beautiful thing we can have, and live freely. We want a life in peace, calm, in community, laboring with each other."*

plot around the arrival of fictional "Kenny Oil" representatives to Shintuya to present exploration plans, based on those actually detailed in the Hunt Oil report. Kenneth Luna, in Peru following his dismissal from the defunct New York Lehman Brothers firm, was typecast to play the visiting Kenny Oil representative, along with his Limeño counterpart, comprising the film's principal comedic duo. The actors gave mock presentations of the report to the community for dramatic effect in the film, while providing an educational forum in Shintuya on details of the real Hunt Oil project impending on their community and life. The plan outlines eighteen seismic lines criss-crossing the reserve with over 20,000 detonation points and 166 heliports carved into pristine jungle.

Enacted scenes created a discursive space in which the Hunt Oil project, an unspoken concern for members of Shintuya, could be openly discussed and publically challenged. As depicted in the film, Valbina Miguel took the opportunity to contradict Kenny Oil's promises of jobs, income, and development with oil exploitation.

> *"You have signed agreements without consulting us, without letting us know what you will do.You are invading four communal reserves, that are protected, that are ours –and this is a lie. You are going to bring contamination, bring diseases, contaminate rivers, fish, and everything is going to die. We don't want you here, go away!"*

Following the scene shoot, a young Shintuya woman, who followed Valbina in expressing her resolve against the oil project told the native members of our cast about the inspirational effect the play exercise had on her by explaining that she had previously never seen an indigenous person speak so forcefully in public. We discovered as the production went on, that filming and playing back the scenes at nightly screenings of the day's work had a marked effect of increasing self-validation and confidence, among indigenous participants, in their own voice and demeanor. We began to commonly refer to this as "camera therapy". The dialogic process of collaborative film workshops succeeded, more broadly, to locally produce the articulation of an alternative indigenous discourse in opposition to the prevalent neoliberal development politics of Alan García's government.

Envisioning Indigenous Protagonism

On the one hand, mock confrontation of multinational oilmen proved important to crafting an image of indigenous resistance to the abuses and absurdities of neoliberal development in the Amazon. However, more crucial was the difficult and nuanced task of depicting the values of indigenous life and the natural world, toward envisioning an Amazonian alternative to development. Such a project necessarily requires the conscious objectification of cultural values through the camera lens. For this task, we collectively chose Brus Rubio, a Bora-Huitoto painter from the northern Peruvian Amazon of Loreto, to portray the heroic figure of the comedy. Through the cinematic vision of Brus, we sought to take up and reverse President Alan García's influential, striking, and discriminatory rhetorical casting of Peru's indigenous peoples as the dog in the Spanish parable "*el perro del hortelano*" (the gardener's dog); the dog who does not eat or allow others to eat, forcing all to go hungry. García has succeeded in framing national discourse on these terms, supplying justification for ongoing neoliberal development of the Amazon (García, 2007). More radically, the film seeks to subtly challenge the presuppositions of "poverty" and "wealth" underpinning the logic of neoliberal development.

Another major hurdle for our production's portrayal of indigenous protagonism, is the stark absence of it in Peruvian cinema, where German director Werner Herzog's Conradian Amazon saga *Fitzcarraldo* (1982), is often considered essential national cinema. With *El Perro del Hortelano*, Brus is the first Amazonian indigenous protagonist in a feature-length Peruvian movie. While placed in Shintuya, the cast of mixed Amazonian groups made the film's indigeneity a national pan-indigenous performance. Brus' portrayal is of an everyman heroic archetype, not ruggedly handsome and bold, rather, light-hearted, deliberate, pensive, and searching. The actions of Brus advancing to uncover and reveal the corrupt and destructive truth of Kenny Oil to Shintuya are intercut with quotidian scenes of Amazonian communitarian life; harvesting and preparing yucca, hunting, familiar jokes, fishing, children playing in the rain, artisanal labor, and the immense natural beauty of the rainforest. The natural and lived landscape of the Amazon is cinematically realized as a counterpoint to the brash cynicism of the oilmen and the NGO volunteers Brus engages with to assist his efforts.

The juxtaposition of crass and manipulative oilmen, with a reserved and passionate indigenous artist, delivers a simple yet powerful moral. Brus triumphs in the film over Kenny Oil, with help from his friends, by hijacking the representatives' final promotional presentation and revealing their real destructive intentions, exhibiting facts of the environmental toil of petroleum exploitation. The message is clear with a chart taken from Hunt Oil's actual environmental impact assessment that lists "economic valuation" of the anticipated damage to the entire natural and human environment such as air, water, and soil pollution in bare dollar amounts.

> "How can they put a price on the elements that give us life? For example, water, they give a price, soil, air they also give a price. This is completely ridiculous! Please, look at what they are doing."

Following Brus' successful counter-presentation, the village children deliver revenge by comically driving out the Kenny Oil representatives into the river.

The other reference for differentiation and development of Brus' character, as well as the film's portrayal of indigenous protagonism, is his ambiguous relationship with Angie, an American researcher who encourages Brus to utilize the resources of the nearby NGO staffed with foreign volunteers in his efforts to organize against Kenny Oil. His relationship with Angie burgeons while Brus uses the NGO's theater program to create protest performances through native myth, and uses its camera equipment to spy on the oilmen and document the malicious use of bribes and threats in the community. Their relationship is the subject of the plot's central reticent romance and reflects the cooperative production's internal contemplations of the ethics and the tensions of intercultural exchange and cooperation.

We drew inspiration for the dialogic method of production from Paulo Freire's classic text, *Pedagogy of the Oppressed.* The basic foundations for cooperative work consisted of building common understanding and themes between indigenous and non-indigenous participants. For instance, Wili Corisepa of Shintuya responded to the workshop discussion I led on the "commodification of nature" with Harakmbut lessons on the "language of nature". However, more critical was the challenging task of collectively recognizing and confronting the present categories and relations of power, the task of reflecting on one's own world in order to change it. In this context, that meant opening to debate the assumptions and meaning of development. The demands of this critical reflexive work created interpersonal tension and conflict

Figure 11.5 Brus speaking to Angie, played by Annika Beaulieu: *"I will put one hundred percent in your projects, but you have to promise me to really open your mind, listen, and understand the indigenous world."*

Figure 11.6 Brus speaking to the community of Shintuya: *"I say to the youth that it is better to stay within the community, and have the voice and rights to lead our people, not abandon it. If we have an anxiety, a fear, it is as a negated voice! Rather, be strong and express what you feel within yourself."*

within our group that made production a constant and uncertain internal struggle between competing voices. This method proved especially challenging for non-indigenous participants, with several participants choosing to leave the project.

Rather than assume one authoritative voice for the narrative, we attempted to reasonably portray different voices and perspectives through the characters' dialogue. Highlights of this internal debate were scripted into an exchange between Brus and Angie over the need for hospitals and schools in the Amazon, questioning whether their value ought to be understood from a perspective of indigenous resistance to impositions of outside knowledge or as objective measures of progress towards improved livelihood.

The moment of greatest intimacy in this relationship is depicted in the film through a romantic candlelit shared reading of the text of Eduardo Galeano's *Open Veins of Latin America*, highlighting the paradoxes of the tortured history buried beneath our contemporary actions and utterances. The scene climaxes with a tepid kiss and a silent moment between Brus and Angie. Their relationship ultimately ends, following Brus' success against Kenny Oil, in disillusionment when he discovers that Angie's research includes collecting data for the US government. For our collaborative production crew, this bitter-sweet ending reflects a relevant message of the ongoing historical weight and reality of colonialism, our intercultural ethics of equability, the limitations of our activism and creative process, and the reality that indigenous peoples' struggle for self-determination is ultimately no one's but their own.

El Perro del Hortelano's Ongoing Story

El Perro del Hortelano has received critical attention in Peru since it premiered at the Amo Amazonia 2009 Film Festival in Lima, Peru. During the post-screening question and answers, we were amazed and inspired to find the dialogic philosophy of the film so readily assumed by the audience at the premiere. Audience members responded by delivering their own impassioned testimonials of experiences with discrimination, racism, and social injustices in Peru. The film has served as a reference point in a renewed national dialogue over development jarred open by the Bagua massacre. Surprisingly, even representatives of the government screened the film during official post-Bagua roundtable talks with the national indigenous Amazonian federation, the Interethnic Organization for the Development of the Peruvian Amazon (AIDESEP). While Peru's commercial film distributors passed up the movie, it received positive national press for achieving a critical cinematic style dubbed "Amazonian Realism" (La República, 2010). The most flattering recognition, however, has come from the Native Federation of the Río Madre de Dios (FENAMAD) which has adopted and promoted the film as its own, as a part of the ongoing struggle against Hunt Oil.

In the months following the film shoot, tensions between FENAMAD, the community of Shintuya, and Hunt Oil escalated to the point of open confrontation. In September of 2009, FENAMAD filed a legal injunction in the local district court in Puerto Maldonado, the regional capital of Madre de Dios, against Hunt Oil's activities and presented Hunt Oil representatives with a declaration rejecting seismic exploration on the reserve. The following month, Hunt Oil's refusal to abide by these demands, claiming their actions were in keeping with the requirements of national law, was met with a mobilization by hundreds of indigenous protesters from Shintuya on Hunt Oil's operational headquarters in Salvación. The protesters threatened to physically expel Hunt Oil if they did not comply with local demands. FENAMAD secretary, Adan Corisepa Neri, framed the protest in terms of the national struggle for realization of indigenous rights, "Laws are being systematically ignored by the company and the government. The Peruvian state has a hydrocarbon policy that violates the rights of indigenous communities. This is what the Amazon uprising was about." (Weinberg, 2009). Met by police, the mobilization ended in disappointment for protesters with the promise of further dialogue between the parties. With little progress made to halt Hunt Oil through either formal legal processes or protest, in 2010, FENAMAD moved into direct political contest, postulating the president of FENAMAD Antonio Iviche Quique for regional governor of Madre de Dios. A Harakmbut native of Shintuya, Iviche Quiche ran under the slogan of regional and indigenous "re-vindication" against corruption. His unsuccessful campaign faced persecution leading to his arrest three weeks prior to the election on state prosecutorial charges, ironically, accused of environmental crimes in illegal mining (Servindi, 2010). State authorities have previously leveled similar charges, later dismissed, against him in both 2008 and 2009 following indigenous led protests in Puerto Maldonado. The criminalization of indigenous politics in the case of Iviche Quiche sharply mirrors the national drama of the persecution of AIDESEP president Alberto Pizango following the Bagua massacre.

El Perro del Hortelano provides a lesson relevant beyond film for cultural production on the importance of methodology. The 2009 Selva Rica film project began production in the Peruvian Amazon with a purely cultural and artistic vision that was in many respects naïve. However, because indigenous groups were active participants of producing the message, we were able to develop a critical project able to both speak to a broad popular audience and radically challenge exclusionary politics and practices of Amazonian development. Through a method of dialogic pedagogy, we fostered a community of intercultural knowledge exchange able to critically reflect on, and actively struggle against, practices of power that threaten the environmental future of the Amazon.

The thematization and depiction of the realities of development in the Peruvian Amazon and indigenous community life, from its general absurdities, paradoxes, subjective contradictions, and quotidian subtleties, that emerged out of this dialogic process gives the narrative its critical weight. The film was able to effectively circulate and inform understanding from the local context of Shintuya to regional and national politics and discourse. Conservation and the environmental future of the Amazon depends on the active and consequential inclusion of voices beyond policy circles, NGOs, and the academy toward a knowledge that can be shared and utilized by the whole community involved, and beyond.

Notes

1 Film credits: Director: Renzo Zanelli Barreto, Producer: Annika Beaulieu, Duration: 89 minutes, from www.perrodelhortelano.com; www.selvarica.com.

2 Best Screen Play — The Latino Environmental Film Festival in Rome, Italy 2010; The Royal Bank of Canada Best Environmental Film Award — Film North Huntsville International Film Festival 2010. Best Narrative Film — United Nations Association of the USA, Global Wake Up Chicago 2011; Best Foreign Film — Southern Appalachian Film Festival 2011.

References

Asociación Interétnica de Desarrollo de la Selva Peruana (AIDESEP), "Carta Abierta al Presidente Alan García." Nov. 13, 2007, from http://www.aidesep.org.pe/editor/documentos/59.pdf.

García, Alan. "El síndrome del perro del hortelano." El Comercio, Oct. 28, 2007, from http://www.aidesep.org.pe/editor/documentos/58.pdf.

La República. "*El Perro del Hortelano* se defiende", Jan. 7, 2010.

Rénique, Gerardo. "Blood at the Blockade: Peru's Indigenous Uprising." North American Congress on Latin America (NACLA), June 8, 2009, from https://nacla.org/node/5879.

Servindi. "Peru: Antonio Iviche asiste a lectura de sentencia." Sept. 15, 2010, from http://servindi.org/actualidad/31483.

Skeen, Lisa. "US Praise for Peru's Economy Misses the Mark." North American Congress on Latin America (NACLA), Sept.13, 2010, from https://nacla.org/node/6739.

Weinberg, Bill. "Peru: Hunt Oil contract to Reignite Amazon Uprising?" North American Congress on Latin America (NACLA), Nov. 9, 2009, from https://nacla.org/node/6231.

12 It's the Glacier's Time to Die

Claes Andreasson

Claes Andreasson was a well-known international journalist, who for our conference set out to see what the conditions and perceptions of climate change were on the ground. On his extensive trip from Northern Peru to Chile and Bolivia, he encountered many troubled voices from people on the street and researchers. Water is a key concern in the minds of highland and lowland indigenous populations.

In 2010, the global emissions of heat-trapping carbon dioxide jumped by the biggest amount on record, according to the U.S. Department of Energy (Blasing and Bodem, 2011). The world pumped about 512 million more metric tons of carbon into the atmosphere than during the previous year, an increase of 6 percent. The levels of greenhouse gases thereby exceed the "worst case scenario" in the 1997 projections from the United Nation's Intergovernmental Panel on Climate Change (IPCC, 2007). The effects of these emissions can already been seen not merely at the Polar Regions, but where 70 percent of the world's people live, as in the tropical parts of South America. People on the ground, in turn, are becoming increasingly aware of the dramatic changes around them and modifying their way of life, eventually to suit much worse conditions.

"Do you see that cirque?" Carlos Olmos asks, pointing to a bowl-shaped depression in the mountain range across the way (National Snow and Ice Data Center, 2011). "That's a sign that we had a glacier there" (Olmos, 2008).

Bolivian hydrologist, Carlos Olmos, has taken me on a tour northwest of the capital, La Paz. After a few hours on narrow and curvy dirt roads, we arrive at *Cordillera Real,* a mountain range where the highest peak, *Illimani* which in the indigenous language *Aymara* means "The Golden Eagle," reaches over 21,000 feet. A few shaggy alpacas leisurely move away as we park. The air is cool and thin. Being unaccustomed to the high altitude, I gasp for oxygen and can only walk a few steps before I need to stop and rest.

"The biggest glaciers are retreating now," Carlos Olmos says. "Since the 1980s we have seen a dramatic change. It cannot be a natural cycle. Natural cycles take thousands of years. This happened in just the past thirty years. I am sure it's climate change." In spite of this rapid disruption, the scenery is breathtakingly beautiful and one of my first stops on a journey as a radio reporter on the lookout for signs of climate change in South America.

On our way back to La Paz, we stop to observe what was once the highest ski resort in the world, at 17,000 feet above sea level, *Chacaltaya*. Its snow-powdered tops are still featured on glossy travel posters, but in reality there is little left of this once majestic glacier.

"*Chacaltaya*, I think, is the best example of the impact of climate change," says glaciologist Edson Ramírez as we meet at the Hydrological Institute in the capital (Ramirez, 2008). He works as the Principal Investigator for the Bolivian Glaciological Program and is concerned about what he sees. "*Chacaltaya* is . . . gone. And I think it is the same for all of the other smaller glaciers in the *Cordillera Real* in Bolivia. They might disappear in the next twenty or thirty years. As for the bigger ones, like the *Illimani*, maybe [will disappear] in the next forty or fifty years."

It is not merely the beauty of these old glaciers that Edson Ramírez mourns. They are also a major source for potable water in La Paz and its sister city El Alto, the combined home to more than 1.5 million people. About 80 percent of the drinking water for these two urban centers comes from rain and melt-off in the highlands. The fact that the distribution system is inferior, and almost half of this precious commodity is lost due to leaking pipes, adds to the challenge.

Figure 12.1 As the glaciers around the altiplano of Bolivia melt, the supply of fresh water for La Paz and El Alto is quickly depleting.

"We can imagine more evaporation in the future and maybe less rainfall, and that is going to be a major problem to access potable water in this area," says Edson Ramírez.

Many other metropolitan areas in South America may face a similar fate. Residents of Lima, Peru are already suffering from water shortages aggravated by climate change. While Chile's capital, Santiago, which also depends heavily upon the melt-off from nearby glaciers, may similarly experience insufficient reserves of potable water in the near future.

As I visit the Chilean Andes in early December, the run-off from a dozen glaciers are gushing down the naked mountainsides. Although it is summer, a chilly wind sweeps through the valley. The rhythmic calls from a goat herder echo across the canyon, as he tries to convince his reluctant livestock to move up the hillside to greener pastures.

It is estimated that about 70 percent of the drinking water in Santiago comes from glaciers like the stately *Echaurren*. "Some people think the *Echaurren* may disappear in only fifty years," says our guide Antonio Vergara, a field worker for the local Water Authority, *Departamento de Agua,* in Santiago (Vergara, 2008). If that prediction is true, it would severely affect people in the capital.

"I think the most alarming sign of climate change is what is happening with the glaciers and the supply of water," says Luis Cifuentes, Professor of Engineering and Head of the Center of the Environment at the Catholic University in Santiago.

Figure 12.2 At the foot of the Copa glacier, sits Copa Grande, a small town that is at the front line of the melting glaciers in the Andes.

> "All our water comes from the Andes. Chile is a very long, but narrow country. Our rivers are short and have very pronounced slopes. To have a consistent supply of water in those rivers, you need a reservoir in the mountains. That reservoir is from the glaciers. When we have precipitation, mostly in the winter, water stays in the mountains as snow and ice. If that doesn't happen, we will have all our water rushing down in the winter, when we don't need it. And we will have no water in the summer." (Cifuentes, 2008)

Eighteen hundred miles to the north of Santiago and 11,000 feet above sea level in the Peruvian Andes, is the small farming village of Copa Grande, home to five hundred families. The village sits at the foot of the Copa glacier, which provides it with water in the dry season. Farmer Marcelino Cruz owns one and a quarter acre, where he grows corn, wheat, beans and potatoes. He has noticed that the rainy season has become more irregular. Sometimes it does not rain for days, sometimes it pours down and erodes his farmlands. The dry season has changed too. "There is much less water from the glacier than there used to be," thirty-four year old Marcelino Cruz says. "When I was a boy, the glacier went all the way down to where you see the waterfall disappearing behind those hills," he says pointing up the hillside. "So it has retreated about two kilometers only in my lifetime," (Cruz, 2009).

The local average yield in Copa Grande is down by at least a third. So far, Marcelino Cruz is trying to make up for it by working a second job. He hopes that his

Figure 12.3 Peruvian farmers have had to readjust their cultivation practices with the changing weather patterns.

two children will grow up to be professionals and get jobs in any of the nearby cities. "These fields will not be able to feed them," he says. When I ask Cruz and other farmers in Copa Grande why they think that the glacier is disappearing so rapidly, many of them answer "it's the glacier's time to die, and when it dies, we will, too."

A warmer climate has also brought changes for the farmers in Valle de Mantaro, a little further south in the Peruvian highlands. In the small farming village of Quilcas, potatoes are a staple. But the farmers I meet are worried. "Everything has changed," says potato farmer Liberata Alcoser Huaman. "It is mid-December and the rainy season should be in full swing, but there can be many days between the rainfalls. We have lost an entire season" (Alcoser, 2008).

To adapt to a changing climate, the *campesinos* of Quilcas have had to start cultivating at higher altitudes. "Potatoes have moved up all over Peru. Amazingly, it is now growing at almost 14,000 feet, which is unheard of," says plant breeder and pathologist Maria Scurrah who runs the non-profit organization *Grupo Yanapai*. "The Quilcas farmers are also selecting their own seed and are trying to find varieties that are better adapted to these new conditions," (Scurrah, 2009).

Maria Scurrah takes me to the potato market in nearby Huancayo. She points to dapper piles of golden *Amarilla Peruana, Huamantanga* and the purple-colored *Paseña* from Huancavelica. "*Yungay* is still the most popular improved variety in Peru," she says. "The native varieties are easy to spot because you can see that the plant breeders go for big, round eyes. Whereas the ancient varieties are the opposite — small, shriveled and with deep eyes."

In spite of the many different kinds of potatoes at the market, this is but a fraction of the more than 3,000 varieties currently grown in the birthplace of the potato. "In recent years, the rainy season starts later in the year, and is more irregular. Frost and heavy hail is more common. And there are new pests attacking the crops," Maria Scurrah says.

"Although we have found some new varieties that seem to be more tolerant, production is down by at least a third," says Quilcas potato farmer Wilber Rodríguez Contreras (2008).

Seventy percent of the world's tropical glaciers can be found in Peru. For a country so dependent upon the melt-off from those glaciers, Peru is likely to see an increasing number of signs of a changing climate. One obvious example is the *Quelccaya* Ice Cap in the southeast of Peru that dates back to 315 A.D. It spans an area about forty-four square kilometers and houses the largest tropical glacier on earth. For more than three decades, Ohio State University glaciologist Lonnie Thompson has studied the glaciers at *Quelccaya* and he has found some remarkable changes. "In the first fifteen years, the glacier was retreating up the valley at about six meters per year. And in the last fifteen years, it has been retreating over sixty meters per year. So, ten times faster," he says. "Since I was a graduate student, we've lost 25 percent of the area of that ice field," (Thompson, 2009).

The future for tropical glaciers, like the ones at *Quelccaya*, is very bleak. As Lonnie Thompson explains:

> A glacier is like a bank account. Typically, you have accumulation of snow and ice at the top and you have ablation, the melting of the glacier, at lower elevations. If the glacier is in balance and is healthy, it will exist through time. But when you start taking out more of this bank account than is going in, eventually the glacier is gone. The

> mean of all our best models of temperature rise by the year 2100 is three degrees Celsius at the surface. When you look at those same projections for the tops of these mountain ranges, where the glaciers are, you are talking about five to six degrees Celsius of warming. Twice as much. And frankly, I think that the glaciers in the tropics are already responding to that warming.

Signs of a changing climate are not merely visible in the highlands of South America. After a couple of days' journey north of *Quelccaya*, I find myself in the Peruvian Amazon region. In a long and narrow *balsa* we cross the *Marañón River* to meet with local *Awajun* leader Simon Wipe Bejus. He is dressed in a beautiful headband of colored beans and feathers, and with two wide belts of red, white and black beans crossing his chest, symbolizing the indigenous peoples' struggle for peace, he says. "The jungle is our market, and nature is our pharmacy. This is where we find our cures and our nourishment," said Simon Wipe Bejus (2009).

We are sitting on the riverbank. It is 38 degrees Celsius and very humid. "The climate is changing. It's not like before. The air is scarce, the rain is scarce and the sun feels three times as hot as it used to be. The mountains are getting dry, and that puts us in danger. The Marañón River used to be brimming with water, but not anymore. The rivers and streams have gotten really small and that really worries us in the Amazon region," he adds.

As the sun reflects off of the swift currents of the Marañon River, it is hard to comprehend that this water will eventually end up in the Atlantic Ocean. The Marañón River is one of the early tributaries of the mighty Amazon River and will carry the melt-off from Peruvian glaciers more than 3,200 kilometers east. In addition

Figure 12.4 Flooding in Belo Monte. Brazil has forced the population to deal with the present effects of global climate change.

to irrigating farmlands, the many tributaries of the Amazon River will also satisfy some of the hunger for more electricity. In Peru, Bolivia, Ecuador and Brazil there are nearly 150 hydro-electric dams in operation, under construction or planned (International Rivers, 2011).

Ironically, the hydro-electric dams emit huge amounts of the greenhouse gases that are causing the retreat of many tropical glaciers! When an area is flooded to create a reservoir, the organic material will rot and thereby release both carbon dioxide and the much more powerful greenhouse gas methane. "You have a big pulse of emissions the first few years that comes from the soft vegetation, the leaves and grasses, and the soil carbon that are there when the forest is flooded. Via photosynthesis the vegetation in the rainforest takes the carbon out of the atmosphere as CO_2 but when the soft vegetation at the bottom of a reservoir rots, it goes back as methane. So you are continually converting CO_2 into methane. It's what I call a methane factory," says Philip Fearnside, Professor of Ecology at the National Institute of Amazon Research, INPA, in Manaus. "Methane has a much higher impact on global warming. The last official number from the Intergovernmental Panel of Climate Change, IPCC, was 25 times per ton of methane as compared to a ton of carbon dioxide" (Fearnside, 2011).

The summer of 2011, Brazil started building what will be the world's third largest hydro-electric dam, Belo Monte, in the state of Pará in eastern Brazil. Along the vivid Xingu River a total capacity of 11,200 megawatts will be installed, while flooding an area of more than 500 square kilometers of the rainforest.

Although there is no official decision yet, it is likely that a second power plant will be built upstream, the Altamira/Babaquara Dam. "For the first ten years, these two dams would average about four times more greenhouse gas emissions than you get with fossil fuels. According to my estimates, they will emit an average of 11.2 million tons of carbon equivalent per year, which is more than the whole city of São Paulo emits," Philip Fearnside says.

The greenhouse gas emissions vary depending upon the size, depth and age of a reservoir. Although producing only 250 megawatts of energy, the Balbina Dam just north of the city of Manaus, is one of the largest reservoirs in tropical South America, covering an area of nearly 1,800 square kilometers. According to a recent study, the Balbina Dam emits the equivalent of three tons of carbon dioxide per produced megawatt of electricity, as much as three coal-fueled power plants (Kemenes, Forsberg, and Melack, 2011).

"By my calculation, it would take 41 years for the Belo Monte and Altamira Dams to break even in terms of global warming. You have a big debt to pay from these huge emissions in the first years, while over time the dams will be producing more electricity as compared to fossil fuel and thereby paying off that year. But it takes 41 years! And it means that dams like these are not "green" energy or clean energy," says Philip Fearnside.

The Belo Monte Dam will not merely flood forested areas, at least five thousand families in the city of Altamira will also be forced to move. As the construction of the site started, more than four hundred families occupied an area of the rainforest on the outskirt of the city. They chopped down the trees, burned them, and started building new homes. "Every rainy season our homes are flooded. Imagine what will happen when the power plant is built. We have to find a new place to live," says Maria Gomes (2011).

While some people cut down the rainforest to build new homes, others slash-and-burn lands to open them for farming and cattle grazing. Driving along the

Figure 12.5 Deforestation in the Amazon presents problems not just for the environment, but for local farmers' livelihoods as well.

Transamazónica Highway east of Altamira, it is obvious that large areas of the rainforest have been cut down. We stop in the small farming village of Belo Jardim, where farmers primarily rely on cattle and cacao. While many growers have deforested sizeable areas in the past, they are now part of a voluntary program to learn better land management. "The last eight years, I have not cut down any of the rainforest on my land," says Orlando Alves Martines (2011). Of his property of fifty hectares, a fourth still has forest standing. He is well aware that he has violated the Brazilian Forestry Code, which requires that at least 80 percent of the property remains forested (World Bank Group, 2010).

His neighbor, Blyts Apelfeler Rais, who has left more than half his property of 200 hectares forested, has learned that slash-and-burn is not good for the climate. They have also been taught how to grow cocoa and corn in the middle of the jungle without cutting down the trees. "People here have changed, we understand that we have to protect the forest. Some years ago, people just wanted to cut and burn. We are also afraid of the increased federal enforcement," says Blyts Apelfeler Rais.

The Amazon is a tremendous carbon sink, soaking up carbon dioxide. It is estimated that the vegetation contains between 90 and 140 billion tons of carbon, equivalent to nine to fourteen years worth of all current human-caused emissions, according to the World Wildlife Foundation (2007). But when the trees are cut down, the carbon is released. The deforestation in all the world's rainforests makes up for about one fifth of all the greenhouse gases in the atmosphere — which is more than the emissions from the entire transportation sector.

After a dramatic decline in Brazil 2004–2010, deforestation in the Amazon jumped sharply the subsequent year. The Brazilian independent research institute Imazon reported an increase of 15 percent (Imazon, 2011), while Brazil's National Institute of Space Research (INPE) recorded a 28 percent rise (National Institute for Space Research 2011). Deforestation is also a major reason why Brazil, in the past twenty years, has nearly doubled its emissions of greenhouse gases (United Nations MDG Indicator, 2011).

Although their total emissions of greenhouse gases are relatively small, deforestation is also a problem in Bolivia and Peru, which is why the Bolivian government has pledged to cut deforestation by half and the Peruvian government has set a very ambitious goal of zero net deforestation in only ten years.

In addition to slash-and-burn and forest degradation, higher global temperatures are also a problem for the rainforest. "When the temperature increases, the water in the soil evaporates and the forest becomes drier. It is putting the place ready for more violent wild fires," says Humberto Gómez with the environmental non-governmental organization *Fundación Amigos de la Naturaleza (FAN)* in Santa Cruz, Bolivia (2008).

Violent swings in the climate, more erratic *El Niño* and *La Niña* events will also affect people living here. In 2008, more than seventy people were killed and over one hundred thousand people were homeless after dramatic storms in Bolivia, according to the country's civil defense authorities. One city that was severely affected was El Torno in southern Bolivia. In early March, more than a hundred families lost their homes after torrential rains flooded the river, swept away houses and caused disastrous mudslides.

"At 5 o'clock in the morning, I suddenly heard something that sounded like thunder. The trees were swaying and after five, ten minutes the mudslide started. Very quickly the whole area was covered in mud," says Marcello Laires (2008).

As we meet a year later, his house is still in shambles and he is living in a home made of a few solid wooden poles covered with blue plastic tarp. The narrow dirt road is lined with thirty-some temporary dwellings. In another part of town, I find Brigitte Candia squatting in front of a big outdoor sink, washing her family's clothes. Her tent has six beds, a small breakfast table and a liquid propane gas stove. There is no electricity in the tent, and the cold water and latrine is shared by all evacuees at the outskirts of the tent camp. "It is very hot inside the tents," Brigitte Candia says, "so we spend most of our time outside. Even in the winter, the children do their homework outdoors" (2008).

No one I meet in El Torno has ever experienced such heavy rains before. Whether the floods were a direct result of climate change or not, there is no doubt that the rainy season now arrives later in the year, and the rains are more intensive and irregular than they used to be. "There is a pattern of flooding over the past years. What we're seeing is that the water levels go higher every year," states Humberto Gómez with FAN. "To live in this new climate, with more heat, drought and flooding, and these catastrophic events happening more frequently — we will have to adapt."

According to data from the Mauna Loa Observatory, the concentration of atmospheric carbon dioxide in July 2011 was 392.39 ppm (parts per million) (CO_2 Now, 2011). The amount of greenhouse gases emitted is steadily increasing. In this past decade, we emitted more than 25 percent more greenhouse gases than we did in the 1990s, and more than twice as much as we did in the 1960s. We have already surpassed the 350 ppm level that many scientists and the Intergovernmental Panel of Climate Change believe to be an "upper safety limit."

It is also sobering to realize that even if we were willing to decrease the emissions we are releasing into the atmosphere, the concentration would still continue to increase. "The lifetime of greenhouse gases is decades to centuries. So even if you cut down the emissions of carbon dioxide by 50 percent in the next 25 years, the concentration will still keep increasing, because they're staying in the atmosphere for such a long time," explains Professor Veerabhadran Ramanathan, Distinguished Professor of Climate and Atmospheric Sciences at Scripps Institution of Oceanography (2010).

Although there are still many unanswered questions when it comes to climate change, it would be fatal to wait until we have a full scientific grasp of the problem, says Terry Chapin, Professor of Ecology at University of Alaska-Fairbanks. "I think the biggest mistake would be to do nothing or to wait until science has more of the answers, because then it will be too late to produce a planet where we would want our children and grandchildren to live" (Chapin, 2008).

There are still people who believe that there is no such thing as climate change, or that burning fossil fuels has anything to do with those changes. Physicist Richard Muller at University of California, Berkeley has long been embraced by the skeptics, primarily for questioning the famous "hockey stick," the temperature graph showing the increased warming of the globe. But in October 2011, the Berkeley Earth Surface Temperature group that he leads released a study essentially supporting earlier temperature compilations. According to this new study, the globe's land surface warmed by 0.911°C over the past half-century, a mere 2 percent less than the estimate from the National Oceanic and Atmospheric Administration (NOAA) (Berkeley Earth, 2011). It has yet to be seen if the new study from the Berkeley group will convince the skeptics, but whether you believe that it is God's will, it is due to a natural cycle, and it is "the glacier's time to die" or that it is a result of man-made greenhouse gas emissions, there is no doubt that there are obvious and very dramatic signs of a changing climate, not only in some remote areas at the poles, but where a majority of the people of the world are living.

Aside from global warming, what my interviewees are keenly aware of is the deep connections between Andean glaciers in the highlands and the continued flow of water in Amazonia's rivers flowing into the Atlantic.

References

Alcoser Huaman, Liberata. Personal Interview. Quilcas, Peru, Dec. 16, 2008.

Alves Martines, Orlando. Personal Interview. Belo Jardim, Brazil, Aug. 9, 2011.

Apelfeler Rais, Blyts. Personal Interview. Belo Jardim, Brazil, Aug. 9, 2011

Berkeley Earth Surface Temperature, Oct. 2011, from http://berkeleyearth.org/.

Blasing, T.J. and Bodem, Tom(2011). "Record High 2010 Global Carbon Dioxide Emissions from Fossil-Fuel Combustion and Cement Manufacture," Carbon Dioxide Information Analysis Center, from http://cdiac.ornl.gov/trends/emis/perlim_2009_2010_estimates.html.

Candia, Brigitte. Personal Interview. El Torno, Bolivia, Dec. 12, 2008.

Chapin, Terry. Personal Interview. Fairbanks, Alaska, U.S., Oct. 14, 2008.

Cifuentes, Luis. Personal Interview. Santiago, Chile, Dec. 3, 2008.

CO_2 Now, 2011, from http://co2now.org/.

Cruz, Marcelino. Personal Interview. Copa Grande, Peru, Oct. 29, 2009.

Fearnside, Philip. Personal Interview. Manaus, Brazil, Aug. 5, 2011.

Gomes, Maria. Personal Interview. Altamira, Brazil, Aug. 6, 2011.

Gomez, Humberto. Personal Interview. Santa Cruz, Bolivia, Dec. 11, 2008.

International Rivers, 2011. "Dams in Amazonia," from http://www.dams-info.org/en#dam_name=&countries%5B%5D=Add+Filter&operational_states%5B%5D=Add+Filter&operational_states%5B%5D=2&operational_states%5B%5D=3&operational_states%5B%5D=1&basins%5B%5D=Add+Filter&capacities%5B%5D=Add+Filter.

Imazon (2011). "Huge Surge in Amazon Deforestation," from http://www.imazon.org.br/news/huge-surge-in-amazon-deforestation.

IPCC Fourth Assessment Report: Climate Change 2007, from http://www.ipcc.ch/publications_and_data/publications_and_data_reports.shtml.

Kemenes, Alexandre, Bruce R. Forsberg and John M. Melack (2011). "CO_2 Emissions From a Tropical Hydroelectric Reservoir (Belbina, Brazil)," Journal of Geophysical Research, Vol. 116, 11. July 21, 2011, from http://www.agu.org/pubs/crossref/2011/2010JG001465.shtml.

Laires, Marcelo. Personal Interview. El Torno, Bolivia, Dec. 12, 2008.

National Institute for Space Research (2011). São José dos Campos, SP, Brazil, fromhttp://www.inpe.br/ingles/news/news_20110817_01.php.

National Snow and Ice Data Center, Boulder, CO, fromhttp://nsidc.org/glaciers/questions/types.html#CIRQUE.

Olmos, Carlos. Personal Interview. *Cordillera Real*, Bolivia, Dec. 8, 2008.

Ramanathan, Veerabhadran. Personal Interview. San Diego, California, U.S., Oct. 2, 2008 and Dec. 2, 2010.

Ramirez, Edson. Personal Interview. La Paz, Bolivia Dec. 8, 2008.

Rodríguez Contreras, Wilber. Personal Interview. Quilcas, Peru, Dec. 16, 2008.

Scurrah, Maria. Personal Interview. Quilcas and Lima, Peru, Dec. 16, 2008 and Nov. 6, 2009.

Thompson, Lonnie. Personal Interview. Columbus, Ohio, U.S., Nov. 13, 2008 and Oct. 13, 2009.

United Nations Millennium Development Goals Indicator, 2011, from http://mdgs.un.org/unsd/mdg/SeriesDetail.aspx?srid=749.

Vergara, Antonio. Personal Interview. Santiago, Chile, Dec. 4, 2008.

Wipe Bejus, Simon. Personal Interview. Yamayaka, Peru, Nov. 3, 2009.

World Bank Group, 2010. "Project Information Document for Rural Environmental Cadastre Technical Assistance Project," from http://www.google.com/url?sa=t&source=web&cd=10&sqi=2&ved=0CFkQFjAJ&url=http%3A%2F%2Fwww-wds.worldbank.org%2Fexternal%2Fdefault%2FWDSContentServer%2FWDSP%2FIB%2F2011%2F01%2F11%2F000334955_20110111230549%2FOriginal%2F579800PID0Init1Box0347745B01PUBLIC1.doc&rct=j&q=Brazil%20Lula%20Presidential%20Decree%20number%207029&ei=dl1hTv3FA63WiAKT17G5Dg&usg=AFQjCNHRgzRPpdPrRm47vCA5m2H92NqUpA&sig2=jFhjAJ4oWB_fKl6UJldE5A.

World Wildlife Foundation, 2007. "Amazon Forests."

Afterword

Ambassador Jeffrey Davidow

Taken together, the chapters in this volume make for depressing reading. This should not be surprising to anyone: the general trends that we have seen in recent decades all point to increasing, and usually overwhelming, challenges to the protection of the environment and of indigenous people in the Amazon as well as in much of the rest of the world.

The inability of the world community (perhaps community is a misleading term in this regard) to confront aggressively global warming is but the most obvious element in the composition of this grim picture. The denialists in the United States and in some other countries have lost the scientific battle, but refuse to concede because in doing so, they would have to acknowledge the needs for solutions which are all, in one way or the other, painful and expensive.

To deny climate change today is equivalent to an individual refusing to accept the need to visit a doctor — in the short run the approach can actually save money, e.g. no medical fees, no treatment costs, etc.. In the long run . . . Well, in the long run, John Maynard Keynes was, of course, correct.

The unwillingness to face scientific reality is a product of an ideologically-motivated fear that such recognition would inevitably lead to more intense government-led intervention in an economic order that is perceived, in this perspective, as functioning best when it functions with little or no government involvement.

The inability to come to grips with the problems of Amazonia have less to do with the fear of government action, than with the ineffectiveness of most of the relevant governments of the region to carry through with any programs that they may design. The weakness of governmental institutions to plan and enforce, to implement fairly, to enact good laws and obey them, to root out corruption, to give voice to the voiceless are all familiar elements in the nations of Amazonia and impact on almost every issue well beyond those dealt with in this volume.

So, is there no hope? Can we not continue to live well without considering the death of indigenous cultures? Should we simply recognize the inevitable and comfort ourselves in the developed world, with the knowledge that we will probably find appropriate methodologies to allow us to adapt to a changing climate and its impact? Could we not simply build a seawall around Manhattan or Miami and start a flourishing pineapple business in Minnesota? We can, and probably we will. The United States has the resources to do so, but the country will be increasingly alone and isolated from much of the rest of the world, which will be desperately searching for water, arable land, or simply revenge.

If there is a bright spot in this dismal picture, it lies in the increasing number of souls truly concerned and willing to search for remedies. Further, the issues can no longer be deprecated by governments as unimportant. The governmental responses might be inadequate or ineffective, but the questions and the growing body of those demanding answers cannot be ignored. The answers, if and when they do come, will have to be based on reality, not ideology. Glib assertions about genocide and the evils of the market will have to be placed to one side, just as the fear of government action and international cooperation will have to be countered. Many of the papers in this volume reveal a healthy realism, and, when all is said, that is the only foundation on which we can continue to seek reasonable responses and policies.

The Editors

James M. Cooper, a Canadian Barrister and Solicitor, has worked on law reform issues in South America more than a decade and a half. In the mid-1990s, his photographic and documentary film work was published in *Marie Claire, The Globe and Mail*, and his documentary work was seen on the British Broadcasting Corporation, Channel Four (U.K.), and CityTV (Canada). As a law professor, he has worked with governments, indigenous groups and non-governmental organizations around the region. He is an Institute Professor of Law and Director of International Legal Studies at California Western School of Law in San Diego.

Christine Hunefeldt, originally from Peru, has traversed the Amazon basin for decades, charting early German scientific explorations and colonization projects. She has published several monographs: Paying the Price of Freedom (1994), analyzing the lives of urban slaves in nineteenth-century Lima; Liberalism in the Bedroom (2000), inquiring about spousal conflicts in the nineteenth century among various ethnic groups. She has also written a general history of Peru (2004, 2010). She is a member of the American Historical Association (AHA) and the Latin American Studies Association (LASA) and has published extensive articles in several languages and in many venues over the past decades. She is a Professor of History and the former Director of the Center for Iberian and Latin American Studies at the University of California, San Diego.

The Contributors

William J. Aceves is Professor of Law and Associate Dean for Academic Affairs at California Western School of Law. Portions of his chapter are based on Professor Aceves's book, *The Anatomy of Torture: A Documentary History of Filártiga v. Peña-Irala* (2007). Professor Aceves participated in several of the lawsuits described in his chapter.

Denise de Alcantara has a Ph.D. in Architecture/Urbanism from the Universidade Federal do Rio de Janeiro (UFRJ). In addition to authoring articles, book chapters and essays, she is currently a Professor in the School of Architecture and Urbanism at the Universidade Federal Rural do Rio de Janeiro (UFRRJ). She has been a Postdoctoral fellow at the Graduate Program in Architecture (UFRJ) from 2010 to 2012, and was a Visiting Scholar and Lecturer at the Center for Iberian and Latin American Studies and Urban Studies and Planning Program, University of California San Diego, and Lecturer at the Center for Latin American Studies, San Diego State University.

Claes Andreasson was a former anchor for a morning news and magazine show on Swedish National Television. He had been a correspondent for Swedish National Public Radio and produced and hosted radio documentaries for the Canadian Broadcast Corporation, and National Public Radio in the United States among others. The Los Angeles Press Club had awarded Claes Andreasson for best radio documentary four times. He was to produce and host the radio, broadcast television version, and podcast series for this project. His untimely death has cut short his involvment in this and many other projects.

Devin Beaulieu is currently a Ph.D. student in Anthropology at the University of California, San Diego. He is a member of the Selva Rica arts cooperative that seeks to promote indigenous and environmental rights and activism through collaborative film projects with local communities in the Amazon and the Andes. His ongoing research focuses on advancing indigenous rights within new plurinational conceptions of the state.

J. Christopher Brown is an Associate Professor of Geography and Environmental Studies, University of Kansas. He is also Director of the Environmental Studies Program. His work spans the areas of ecology and biogeography, political ecology, and moral geography. Current projects include studies on environmental governance along the mechanized agricultural frontier of Amazonia; social movements, civil society, and the environment; and farmers' land-use decisions *vis-à-vis* changes in the bio-fuel economy in Brazil and the US Great Plains. Among his publications are articles in the *Journal of Biogeography*, *Ambio*, *Comparative Political Studies*, *Latin American Research Review*, *Political Geography*, *Progress in Development Studies*, *Professional Geographer*, and *Geoforum*.

Jeffrey Davidow served as a U.S. Foreign Service Officer in the Department of State for 34 years, retiring with the rank of Career Ambassador in 2003 after having served as ambassador to Zambia, Venezuela, and Mexico and as Assistant Secretary of State for the Western Hemisphere. After retiring from the State Department, he served for eight years as the President of the Institute of the Americas, a not-for-profit center for the promotion of western hemisphere development. He is currently a Senior Counselor of The Cohen Group, a Washington-based organization headed by former Secretary of Defense William Cohen, which provides strategic governmental affairs advice to international businesses.

Richard Finkmoore is Professor of Law at California Western School of Law specializing in environmental and natural resources law. He is a graduate of Stanford University and Stanford Law School, where he was publications director of the Environmental Law Society and interned with the Natural Resources Defense Council. Following graduation, Professor Finkmoore clerked for a justice of the Washington Supreme Court, served as an assistant attorney general for the State of Washington, and was a partner in a law firm handling a variety of civil matters. He is the author of *Environmental Law and the Values of Nature* (published in 2010) and editor of *The Natural Resources Law Manual* (published by the American Bar Association). Professor Finkmoore taught for eight years as a visiting professor in the Muir College Interdisciplinary Studies Program at the University of California, San Diego, and has taught international environmental law in four law school foreign study programs and property law at the National Judicial College.

Lawrence A. Herzog is Professor and Chair of the Graduate Program in City Planning, School of Public Affairs, San Diego State University, San Diego, California. Herzog has written or edited 8 books, including: *Global Crossroads: Planning and Infrastructure for the California-Baja California Border Region* (San Diego: Trans Border Institute, University of San Diego, 2009); and *Return to the Center: Culture, Public Space and City-Building in a Global Era* (University of Texas Press, 2006). Awards include: Fulbright Scholar (Peru), Senior Fulbright Specialist (United Kingdom), Outstanding International Scholar Award (SDSU), CSU International Fellow (University of Paris). He has been an urban/regional planning consultant to national and international agencies such as the U.S. Agency for International Development, the U.S. Department of Housing and Urban Development, and the California Public Policy Institute.

Matthew T. Koeppe is a member of the Adjunct Faculty in the Department of Geography at George Washington University. His research focuses on human-environment relationships in Latin America, particularly in the Brazilian Amazon. Of particular interest are global commodity chains, food geographies, and the social aspects of land-use and land-cover change.

Rodrigo Montoya is an Emeritus Professor of Anthropology at the Universidad Nacional Mayor de San Marcos in Lima, Perú, and the author of many books and articles. As a well-known anthropologist, his research over the past decades has revolved around the living conditions of Andean peasants, their culture, their music, their fight over land and citizen rights. He has also engaged in extensive research in the Amazon Basin and has been an advisor to indigenous organizations both in Peru and in Brazil. He has also been a visiting professor at universities in Spain, Chile, Bolivia, Ecuador, Brazil, France, and the United States. His most recent books include *Nunca más sin nosotros*, *El futuro de la cultura Quechua*, and an extensive essay on the life and work of Peruvian writer José Maria Arguedas.

Nancy Postero is an Associate Professor in the Anthropology department at the University of California, San Diego. Formerly a human rights lawyer and a radio journalist, she received her Ph.D. from University of California, Berkeley in 2001. Her work focuses on the intersection of race, politics, and political economy. She has carried out fieldwork with the Guaraní people of lowland Bolivia since1994. Along with numerous articles on indigenous politics in Bolivia, she is the author of *Now We Are Citizens, Indigenous Politics in Post-multicultural Bolivia* (Stanford University Press, 2007); the co-editor with Leon Zamosc of *The Struggle for Indigenous Rights in Latin America* (Sussex Academic Press, 2003); and the co-editor with Philip Oxhorn of the special issue of Latin American Research Review on *Actually Existing Democracies* (2010).

Anna Sartorio is receiving her Master's Degree in "Environment and Development" at the Universidad Simón Bolívar in Caracas, Venezuela. She is a founding member of the NGOs "AVVA Frontera Gran Sabana" and "Mining Watch for Latin America" in Venezuela and has served as an advisor to the Committee on Environment and Regional Planning of the Senate of Venezuela (1994–1998) and to the Committee on Mining of the House of Representatives (1997–1998).

Stefano Varese, a Peruvian anthropologist of Italian descent, is Professor Emeritus of Native American Studies at the University of California, Davis. He is the author of numerous works including: *Salt of the Mountain: Campa Ashaninka History and Resistance in the Peruvian Jungle*; *Witness to Sovereignty; Essays on the Indian Movement in Latin America*; *La Ruta Mixteca*; *El Impacto*

Etnopolíticode la Migración Transnacional de los Pueblos Indígenas de México (Co-Editor with Silvia Escárcega); and *Pueblos Indios, Soberanía y Globalismo.*

Carlos H. Waisman obtained his Ph.D. in Sociology at Harvard University. He is a Professor of Sociology and International Studies at the University of California, San Diego. His work has focused on the comparative analysis of the incorporation of the working class into the political system, the social and political consequences of protectionism and free trade, and the comparative analysis of democratization and institutional design in Latin America and Eastern Europe. Much of his work has dealt with the Argentine case in a comparative perspective. He has published, among other works: *Reversal of Development in Argentina: Postwar Counterrevolutionary Policies and their Structural Consequences* (1987); *Institutional Design in New Democracies: The Cases of Eastern Euroope and Latin America* (co-edited with Arend Lijphart, 1996); *Civil Society in Latin America* (with Richard Feinberg and Leon Zamosc, 2006) and a large number of articles and book chapters.

Edgard Yerena is a biologist and lawyer who has held the following positions: planning officer of the National Parks Institute of Venezuela (1985–1993); advisor to the Committee on Environment and Regional Planning of the Senate of Venezuela (1994–2000); and consultant for several environmental NGOs. Yerena currently heads the Department of Environmental Studies at the Universidad Simón Bolívar in Caracas, Venezuela.

Index

Page numbers in italics refer to figures or tables.